AF773975

Aquatic Plants

Aquatic Plants

Edited by:

Akansha Singh

Delve
PUBLISHING

www.delvepublishing.com

Aquatic Plants

Akansha Singh

Delve Publishing
224 Shoreacres Road
Burlington, ON L7L 2H2
Canada
www.delvepublishing.com

Email: orders@arclereducation.com

© 2022 Delve Publishing

ISBN: 978-1-77469-169-4 (Hardcover)

This book contains information obtained from highly regarded resources. Reprinted material sources are indicated and copyright remains with the original owners. Copyright for images and other graphics remains with the original owners as indicated. A Wide variety of references are listed. Reasonable efforts have been made to publish reliable data. Authors or Editors or Publishers are not responsible for the accuracy of the information in the published chapters or consequences of their use. The publisher assumes no responsibility for any damage or grievance to the persons or property arising out of the use of any materials, instructions, methods or thoughts in the book. The authors or editors and the publisher have attempted to trace the copyright holders of all material reproduced in this publication and apologize to copyright holders if permission has not been obtained. If any copyright holder has not been acknowledged, please write to us so we may rectify.

Notice: Registered trademark of products or corporate names are used only for explanation and identification without intent of infringement.

Delve Publishing publishes wide variety of books and eBooks. For more information about Delve Publishing and its products, visit our website at www.delvepublishing.com.

ABOUT THE EDITOR

Dr. Akansha Singh, is presently working as Associate Professor & Head, Department of Genetics and Plant Breeding, College of Agriculture, Parul University, India. She graduated from the Banaras Hindu University, with doctorate in Genetics and Plant breeding in the year 2012. She has also worked as Research Scientist in Bioved Research Society, Allahabad, India. She was working as Postdoctoral fellow in University of Mumbai before joining her current position. She has authored a number of national and international publications.

TABLE OF CONTENTS

LIST OF FIGURES

LIST OF ABBREVIATIONS

ACE	Angiotensin Converting Enzyme
ASP	Amnesic Shellfish Poisoning
ATP	Adenosine 5'-Triphosphate
Aβ	Amyloid Beta
BOD	Biological Oxygen Demand
CCM	Cerebral Cavernous Malformations
CFCs	Chloro-Fluoro-Carbons
CFP	Ciguatera Fish Poisoning
COD	Compound Oxygen Interest
COX 2	Cyclooxygenase-2
DHA	Docosahexaenoic Acid
DIC	Dissolved Inorganic Carbon
DNA	Deoxyribonucleic Acid
DPA	Docosa-Pentaenoic Acid
DSP	Diarrheic Shellfish Poisoning
EER	Energy Efficiency Ratio
EPA	Eicosatetraenoic Acid
GRAS	Generally Recognized as Safe
HIV	Human Immunodeficiency Viruses
HRAP	High-Rate Algal Ponds
IL	Interleukin
LCP	Light Compensation Point
LPS	Lipopolysaccharide
MDGs	Millennium Development Goals
NOS	Nitric Oxide Synthase
OF	Overflow
P&Z	Planning and Zoning

PAR	Photosynthetically Active Radiation
PBRs	Photo-Bio-Reactors
PCE	Personal Consumption Expenditures
PCHO	Polysaccharide
POC	Proof of Concept
POTW	Publicly Owned Treatment Works
PUFA	Polyunsaturated Fatty Acid
RAM	Root Apical Meristem
RFS	Renewable Fuel Standard
RH	Relative Humidity
RI	Rapid Infiltration
ROS	Reactive Oxygen Species
SAM	Shoot Apical Meristem
SDA	Stearidonic Acid
SPS	Sulfated Polysaccharides
SR	Slow Rate
STEs	Sieve Tube Elements
TEPs	Transparent Exopolymer Particles
TNF-α	Tumor Necrosis Factor-Alpha
UV	Ultraviolet
UVBR	Ultra Violet-B Radiation
UVR	Ultra-Violet Radiation
WH	Water Hyacinth

GLOSSARY

A

Acropleustophytes – large plant that floats freely at the surface of the water

Alkalinity – is the capacity of water to resist changes in pH that would make the water more acidic

Allelopathy – is a biological phenomenon by which an organism produces one or more biochemicals that influence the germination, growth, survival, and reproduction of other organisms

Amphibious – relating to, living in, or suited for both land and water

Amphibious – able to live on both land and in water

Anatomy – the branch of science concerned with the bodily structure of humans, animals, and other living organisms, especially as revealed by dissection and the separation of parts

Anemophily – pollination done through wind

Angiosperms – also known as Angiospermae, or Magnoliophyta, are the most diverse group of land plants, with 64 orders, 416 families, approximately 13,000 known genera and 300,000 known species

Anoxic – area of seawater, freshwater, or groundwater that are depleted of dissolved oxygen and are a more severe condition of hypoxia

Aquaculture – also known as aquafarming, is the farming of fish, crustaceans, mollusks, aquatic plants, algae, and other organisms

Arachidonic Acids – is a polyunsaturated omega-6 fatty acid 20:4, or 20:4

Aragonite – is a carbonate mineral, one of the three most common naturally occurring crystal forms of calcium carbonate, $CaCO_3$

Azolla – is a genus of seven species of aquatic ferns in the family Salviniaceae

B

Biorefineries – is a refinery that converts biomass to energy and other beneficial byproducts

Biotechnology – is a broad area of biology, involving the use of living systems and organisms to develop or make products

C

Calcification – is the accumulation of calcium salts in a body tissue. It normally occurs in the formation of bone, but calcium can be deposited abnormally in soft tissue, causing it to harden

Carotene – The term carotene is used for many related unsaturated hydrocarbon substances having the formula $C_{40}H_x$, which are synthesized by plants but in general cannot be made by animals

Carotenoids – also called tetraterpenoids, are yellow, orange, and red organic pigments that are produced by plants and algae, as well as several bacteria, and fungi

Cattail – are upright perennial plants that emerge from creeping rhizomes. The long tapering leaves have smooth margins and are somewhat spongy

Cellulose – an insoluble substance which is the main constituent of plant cell walls and of vegetable fibers such as cotton

Cellulose – is an organic compound with the formula $_n$, a polysaccharide consisting of a linear chain of several hundred to many thousands of β linked D-glucose units

Chlorofluorocarbons – chlorofluorocarbons and hydrochlorofluorocarbons are fully or partly halogenated paraffin hydrocarbons that contain only carbon, hydrogen, chlorine, and fluorine, produced as volatile derivative of methane, ethane, and propane

Chlorophyll – the green substance in plants that takes in light from the sun to help them grow

Contamination – is the presence of a constituent, impurity, or some other undesirable element that spoils, corrupts, infects, makes unfit, or makes inferior a material, physical body, natural environment, workplace, etc.

Crayfish – are freshwater crustaceans resembling small lobsters (to which they are related)

Cyanobacterium – also known as Cyanophyta, are a phylum consisting of free-living photosynthetic bacteria and the endosymbiotic plastids, a sister group to Gloeomargarita, that are present in some eukaryotes

D

Debris – is rubble, wreckage, ruins, litter and discarded garbage/refuse/trash, scattered remains of something destroyed, discarded, or as in geology, large rock fragments left by a melting glacier etc.

Dicotyledonous – also known as dicots, are one of the two groups into which all the flowering plants or angiosperms were formerly divided

Docosahexaenoic Acid – is an omega-3 fatty acid that is a primary structural component of the human brain, cerebral cortex, skin, and retina

Duckweed – is a subfamily of flowering aquatic plants, known as duckweeds, water lentils, or water lenses

Dunaliella – is a single-celled, photosynthetic green alga, that is characteristic for its ability to outcompete other organisms and thrive in hypersaline environments

E

Ecosystem – a biological community of interacting organisms and their physical environment

Egeria Densa – the large-flowered waterweed or Brazilian waterweed, is a species of Egeria native to warm temperate South America in southeastern Brazil, Argentina, and Uruguay

Eichhornia – commonly called water hyacinths, was a polyphyletic genus of the aquatic flowering plants family Pontederiaceae

Eicosatetraenoic Acid – is an omega-3 fatty acid. In physiological literature, it is given the name 20:5

Embolism – also called thromboembolism, is a blockage in one of the arteries of the body due to a blood clot that has broken off from another location in the body (embolus) and traveled through the bloodstream to lodge in a small blood vessel

Entomophily – form of pollination in flowering plants through insects

Epiphytes – is an organism that grows on the surface of a plant and derives its moisture and nutrients from the air, rain, water or from debris accumulating around it

Equilibrated – bring into or keep in equilibrium

Ethanol – Ethanol is an organic chemical compound. It is a simple alcohol with the chemical formula C_2H_6O

Eukaryotes – are organisms whose cells have a nucleus enclosed within a nuclear envelope

F

Fertilizers – is any material of natural or synthetic origin that is applied to soil or to plant tissues to supply plant nutrients

Filamentous – filament-like or thin

Foliage – Plant leaves collectively

G

Geology – is the study of the Earth, the materials of which it is made, the structure of those materials, and the processes acting upon them

H

Habitats – in ecology, the term habitat summarizes the array of resources, physical and biotic factors that are present in an area, such as to support the survival and reproduction of a particular species

Hyacinth – A bulbous plant of lily family having straplike leaves and a bell-shaped fragrant flower

Hydrophily – uncommon form of pollination, pollen distributed through flow of water

Hydrophytes – plant that grows only in or on water

M

Macrophytes – type of aquatic plant that grows near water either emergent, submergent or floating

Malnutrition – is a condition that results from eating a diet in which one or more nutrients are either not enough or are too much such that the diet causes health problems

Meristematic Tissue – meristematic tissues have undifferentiated cells which form the building blocks of the specialized plant structures

Microorganisms – or microbe, is a microscopic organism, which may exist in its single-celled form or a colony of cells

Molecule – A molecule is an electrically neutral group of two or more atoms held together by chemical bonds

N

Nutraceutical – or 'bioceutical' is a pharmaceutical alternative which claims physiological benefits

Nutrients – are compounds in foods essential to life and health, providing us with energy, the building blocks for repair and growth and substances necessary to regulate chemical processes

O

Osmosis – movement of a solvent (such as water) through a semipermeable membrane (as of a living cell) into a solution of higher solute concentration that tends to equalize the concentrations of solute on the two sides of the membrane

P

Paleoherbs – the paleoherbs are a small group of flowering plants which have traditionally been classified as dicots

Parasites – parasites include various protozoa and worms that may infect humans that cause parasitic diseases

Pathogens – a pathogen in the oldest and broadest sense, is anything that can produce disease

Pennywort – commonly known as Gotu Kola, Brahmi, Indian pennywort and Asiatic pennywort, is an herbaceous, perennial plant in the flowering plant family Apiaceae

Philoxeroides – commonly referred to as alligator weed, is a native species to the temperate regions of South America, which includes Argentina, Brazil, Paraguay, and Uruguay. Argentina alone hosts around 27 species that fall within the range of the genus Alternanthera

Phloem – is the living tissue in vascular plants that transports the soluble organic compounds made during photosynthesis and known as photosynthates, in particular the sugar sucrose, to parts of the plant where needed

Photobioreactor – A photobioreactor is a bioreactor that utilizes a light source to cultivate phototrophic microorganisms

Photosynthates – is the resulting product of photosynthesis, these products are generally sugars

Photosynthesis – s a process used by plants and other organisms to convert light energy into chemical energy that can later be released to fuel the organisms' activities

Physiology – study of the functioning of living organisms, animal or plant, and of the functioning of their constituent tissues or cells

Phytoplankton – planktons that consist of microscopic plants

Phytoremediation – is a bioremediation process that uses various types of plants to remove, transfer, stabilize, and/or destroy contaminants in the soil and groundwater

Pigment – A pigment is a colored material that is completely or nearly insoluble in water

Pollination – the act of transferring pollen grains from the male anther of a flower to the female stigma

Potamogetonaceae – commonly referred to as the pondweed family, is an aquatic family of monocotyledonous flowering plants

Propagule – vegetative structure that can become detached from a plant and gives rise to a new plant

R

Rhizosphere – is the narrow region of soil or substrate that is directly influenced by root secretions and associated soil microorganisms known as the root microbiome

S

Spirulina – is a biomass of cyanobacteria that can be consumed by humans and animals

T

Taxonomic affinities – refers to the resemblance suggesting a common descent, phylogenetic relationship or type.

V

Verticil – a whorl or verticil is an arrangement of leaves, sepals, petals, stamens, or carpels that radiate from a single point and surround or wrap around the stem or stalk

W

Water Milfoil – also known as Myriophyllum, is a genus of about 69 species of freshwater aquatic plants, with a cosmopolitan distribution

Wetlands – is a distinct ecosystem that is flooded by water, either permanently or seasonally, where oxygen-free processes prevail

X

Xylem – is one of the two types of transport tissue in vascular plants, the other being phloem

PREFACE

This book takes the readers through several aspects of the aquatic plants. This book sheds light on the biology of the aquatic plants, advances in biofuel production with the help of algae, marine microalgae in food and health industry, ecology and management of the aquatic plants, various organ and the functions of water plants, and aquatic plants for water treatment and resource recovery.

The first chapter stresses on the basic overview and introduction to the aquatic plants so that the readers are clear about the philosophies behind that form the utmost basics in the field. This chapter will also emphasize the history of aquatic plants, the life of aquatic plants, different types and uses of aquatic plants.

The second chapter takes the readers through the concepts of biology of aquatic plants. This chapter will provide highlights on the anatomy of aquatic plants, anatomical adaptations of aquatic plants, life forms of the aquatic plants, evolution, distribution and diversity of aquatic plants, cellular biology, physiology, and morphological adaptions of the aquatic plants.

Then, the third chapter explains the advances in biofuel production with the help of algae and various aquatic plants. It also explains the benefits of the microalgal based biofuels, various challenges for algae fuel commercialization, algae-based bioenergy products, technological opportunities and threats for developing countries in algae-based biofuel products.

The fourth chapter introduces the readers to the philosophy behind the marine and microalgae in the food and health industry. This chapter also explains the importance and current uses of microalgae as food, use of microalgae and cyanobacteria for health applications. Antiviral properties of marine algae, health benefit of microalgae in preventing cardiovascular diseases, along with its future prospects.

The fifth chapter throws light on the ecology and management of aquatic plants. This chapter explains free-floating plants, macrophytes, biotic interactions, the importance of aquatic plant management and its various types, control methods for aquatic plants in ponds and lakes, along with various aquatic plant control methods.

The sixth chapter takes the readers through the significance of various organs and their functions in water plants. The readers are then told about the plant body, concept of meristems, genetic control of flowers, vegetative parts of the plant, various types of roots, stems and their structure, movement of water and minerals, and nutritional requirements of the plants.

The seventh chapter explains the application of aquatic plants for water treatment and resource recovery. This chapter also emphasizes the wastewater treatment system, utilization of the aquatic plants to treat wastewater, artificial wetlands, and cultivation of aquatic plants for biofiltration of wastewater.

The last chapter of this book sheds light on the future economic and environmental impacts of microalgal technology. This chapter also mentions the impact of changing global environment on the water plants, environmental impacts of the microalgae technology, impacts to aquatic biodiversity, and terrestrial impacts.

This book has been designed to suit the knowledge and pursuit of the researcher and scholars and to empower them with various aspects of the aquatic plants so that they are updated with the information. I hope that the readers find the book explanatory and insightful and that this book is referred by the scholars across various fields.

Chapter

1

An Introduction to Aquatic Plants

CONTENTS

This chapter deals with the life of plants that live under the water or float on the water. They differ in their qualities and hence are classified into various types. From the primitive age to the recent years, this chapter gives an insight into the life process of aquatic plants. Their life forms, the conditions that they adapt and are suitable to them.

It discusses the different types of aquatic plants, along with their use in different fields like the food, agriculture, harvesting, and many other uses. It also shows how aquatic plants are useful in economic aspect. Towards the end it also discusses and relates aquatic plants to the ecosystem and their effects, influence on each other.

1.1. INTRODUCTION

Aquatic plants are the plants that grow in water bodies which can be freshwater, saltwater, and brackish water. They have adapted to live in the water like ponds, lakes, rivers, and the sea. There is a special type of requirement that aquatic plants have for living in submerged water or at the surface of the water. Some aquatic plants are completely submerged, whereas some are only partly submerged.

The very usual adaptation which is shown by these plants are the leaves that float on the water and special tissue which permits to transport air and gases inside the plant. Some plants from the water are adapted in land. Some of these plants float on water while others live underwater.

They are further divided into three parts, they are as follows: fixed plant, floating plant, and underwater plants. Explaining the fixed plant, they are the plants that have fixed roots, which are attached to the soil present under the water.

Figure 1.1: An illustration of aquatic plants.

Source: Image by Public Domain Pictures.

They have long, thin and hollow stems in order to reach the surface of the water. Leaves of these plants are broad and flat, it also has a waxy covering on the surface which prevents them from rotting. The examples for these plants are water lily and lotus. Secondly, floating plants are the plants having stems that are light and spongy which floats on the water, whereas the leaves have stomata on the upper side.

Duckweed, water lettuce and water hyacinth are the examples of floating plants. Underwater plants are the plants that grow under the surface of water and have thin and narrow leaves without having any stomata. The leaves absorb carbon dioxide and oxygen directly through their surface, for instance, hydrilla and tape grass.

Aquatic plants, are often referred to as macrophytes or hydrophytes are an important functional constituent of many lakes and reservoirs (Carpenter and Lodge, 1986). These terms are commonly used for the vascular plants, mosses, and larger forms of algae such as filamentous algae and the stonewort – Characeae, which can be seen without the use of any instruments although smaller algae and phytoplankton are eliminated from this.

The vascular aquatic plants are not developed in water rather they are a descendant of terrestrial plants. Fundamentally they manifest physiological resemblance up to a great extent in respect to their terrestrial forbearers but depict some adaptation to under water and saturated conditions as stated by Sculthorpe in 1967.

Macrophytes gets braced by almost all the water bodies, and where they happen, these plants prove to be a vital source of habitat and food for other organisms and perform a key function in nutrient cycles and gaseous exchange through photosynthesis.

1.2. HISTORY OF AQUATIC PLANTS

The history of land plants is traced from aquatic ancestors. Meaning to say, they are originated from aquatic plants. Species from approximately 100 flowering plant families, together with some ferns, mosses, and liverworts, have intruded and started their life in the water.

Most land plants can cope up with flooding for a short duration of time. So, an accurate definition for aquatic plants does not actually prevail. As stated earlier, the term is generally used for plants that grow completely underwater or with leaves floating on the surface.

Segment of the shoot, specifically the ones known as the flowering stems, will at times grow up above the surface of the water. Nearly all aquatic plants live in freshwater such as in lakes, ponds, reservoirs, canals, or rivers and brook. These habitats act very differently within the deepness of water, flow rates, temperature, acidity and alkalinity, i.e., the pH and mineral content.

Figure 1.2: An illustration of water lily plant.

Source: Image by Pixabay.

The living place of some aquatic plants are in river estuary, with a constantly changing blend of fresh water and saltwater, and a few of them live fully under and deep inside the sea. One of the finest examples of such plants are seagrasses. As mentioned above, aquatic plants can be free-floating such as water hyacinths or another type is they are embedded to the base of the pond or stream like water lilies.

The most vital grain crop across the world that is rice, is also an aquatic plant. Rice plants are considered to be semi-aquatic in nature and are semi-aquatic grasses. The height of rice plant varies from 0.4 m to 5 m in the context of floating.

Generally, the main concern for aquatic plants is not water supply. They do not require water resistant cuticles or a large amount of woody tissue to keep them upright. Nevertheless, the life of plants growing underwater is a great challenge. The most important requirement of all green plants is that they need oxygen and carbon di oxide.

These gases cannot diffuse easily and rapidly from the air down through the water. There are many changes made in physiology and in the structure in water plants are the modification and adjustments to find out the solutions regarding these problems related to the gas exchange. Most of the water plants evolve huge internal spaces called the parenchymatous tissue.

These tissues generate in the roots and shoots of the plants. The available air spaces make the tissues buoyant and assists them to float on the surface. The amount of oxygen which is released by the green leaves, photosynthesizing underwater in the light can be stored for a period of time, although not for a very long duration in the air space and then can be made use in the cellular respiration.

Oxygen diffuses down easily through the air spaces when the plant grows or floats to the surface. Nuphar, one of the water lilies, has generated a system of ventilation in order to circulate air from the leaves floating on the surface of water down to its roots in the mud.

The fast increase of shoots to the surface known as the depth accommodation growth which is evident in rice seedlings and countless other kinds of aquatic and amphibious, is operated by the force of the shoot and an expansion of carbon dioxide and the gas that are produced naturally like ethylene, in the tissues submerged. The plant organs present in the mud at the bottom of lakes continue to exist for longer periods without an ample amount of oxygen.

The basic need for aquatic plants is carbon dioxide for the process of photosynthesis. The pH and temperature determine the quantity of carbon di oxide which is diffused into the water. Aquatic plants are sometimes responsible for acidifying their leaf surfaces.

Due to which the carbon dioxide is released from carbonate and bicarbonate salts that are dissolved in the water. Underwater leaves are split up very well, with a large surface area. Dissimilar to land plants, they might contain chloroplasts for photosynthesis in the exterior cell layer called the epidermis.

Figure 1.3: Picture depicting an illustration of Nuphar variegatum.

Source: Image by Wikimedia Commons.

The amount of light obtainable for photosynthesis decreases with depth mainly in unclean water, and the quality of light, the portion of red, far-red, and blue light is also transformed. Some species can produce underwater, floating, or emergent leaves, each with a variety of shapes and structures.

These all are carried on upon on the same plant, and this occurrence is called heterophylly. This is a retaliation to light quality and the quantity of a plant hormone, abscisic acid, which is at moderately elevated levels in emergent shoots. The examples of such species are Potamogeton and Sagittaria.

Most of the aquatic plants give flower and set seed. Although a lot of them have the capability of growing instantly and can also reproduce vegetatively. Aquatic plants, especially the free-floating species, can occupy the surface of a water body very readily.

In case if the water gets parched, then plants have the ability of producing a different kinds of tubers and resting buds, i.e., turions that will continue to remain in the mud until the water comes back and the dryness is gone. These qualities make aquatic plants some of the most troublesome and persistent weeds in the world, mainly in tropical and subtropical countries.

There is, reasonably, a significant amount of unwillingness to set down herbicides into rivers and lakes. A large number of control methods, including mechanical harvesting and the foundation set for fish and other animals to eat them. Techniques like these have been adapted and methods ventured.

In late nineteenth-century Britain, the canal system was obstructed with introduced Elodea canadensis, the water hyacinth, *Eichhornia crassipes* outspread from tropical South America towards the waterways and the canals of Africa, Asia and North America.

Salvinia auriculata, which is a free-floating aquatic fern enveloped quickly the lake, which was of 190 kilometers, beyond the Kariba Dam on the Zambezi River in Africa in the 1960s. The waterways and everglades of the south eastern United States have been time and again engulfed by the unfamiliar aquatic plants which spread rapidly.

Most of the aquatic habitats and the aquatic plants that reside and grow in them are therefore under permanent danger from pollution and from drainage and waste for urban and industrial development all over the world.

Aquatic plants are important factors that help in extracting the pollutants from industries and cleanse the water supplies. They are also an essential part of an aquatic ecosystem that is operative, effective and fully functional for fish and other wildlife.

Figure 1.4: An illustration of Potamogeton perfoliatus.

Source: Image by Wikimedia Commons.

In order to determine that a lake or a stream will be a good producer of fish or not in that case the presence of aquatic plants is one of the finest indicators. The food that fishes require to survive are the aquatic plants. Plants also provide a fishery with protective cover and life-giving oxygen.

Plants are in dire need of sunlight for their growth and photosynthesis to take place. Plants are limited to waters receiving sunlight. Mostly water have a depth of 10 to 30-feet which is the limit to which sunlight can pass through.

This means that plenty of plants are present in the shallows and reduces in quantity as the water intensifies and escalates. Shallow shorelines that remain conserved from floods and wave actions advocates the most abundant growths.

More often than not, there are three dimensions of aquatic plants. The first zone includes mainly rooted plants with their tops elongated to the surface air. These consist of emergent plants. They transpire in depths from the border of shorelines to about the six-foot depth. Cattails, reeds, and others are the plants that are usually found in this first zone.

Talking about the second zone, it is made up of rooted plants which covers floating portions on the surface. This zone is deeper as compared to the first zone and extends to about the 10-foot depth. Common examples are Waterlilies and pond weeds.

Shifting the focus to the third zone, it lies in the deeper areas further the first two zones. It carries submerged plants like eelgrass, Anarchasis and a few others. This zone extends closer to a 10-foot depth. In very deep, crystal clear waters where sunlight can deeply pass through, this zone might advance to about the 30-foot depth.

Immense dense growths of engulfed plants are frequented by feeding fish. Submerged plants in the deeper areas are not clearly visible to the eyes, and it is also difficult to direct their presence. To a moderate extent these three zones converge. In areas of heavy currents, flood and wave action can deteriorate the growth of the plant, and in some of these areas' plants may be missing.

The most beneficial plants to discover are present in the zone three that have submerged growths. These hold up sufficient amount of depth for protective cover and an abundant supply of food.

Different aquatic life forms approbate particular plants. For instance, Callabaetis mayflies, dragonflies and damselflies are to be found in specific plants such as the ones possessing submerged varieties. These insects are stalkers and wait to devour upon their victims. Fish forage the aquatic plants looking for food are the plants responsible for generating a large amount of food supply of the fishes.

1.3. LIFE OF AQUATIC PLANTS

Aquatic plants are the kind of plants that are referred to as true plants or the members of kingdom Plantae. These plants are capable of flourishing and thriving on their own and are also responsible for completing their life cycle while being present in water or on the surface of water or also on hydric soils.

These hydric soils evolve and takes place when the ground is flooded or swamped for a long period of time during the course of the season of growing in order to become anaerobic, that is the depletion of oxygen in the rooting zone.

Meaning to say, this depletion of oxygen occurs in the aquatic environments as the dissolved oxygen becomes reduced in the concentration to a point where it proves to be harmful to aquatic organisms living and surviving in the system.

These soils incorporate organic that is peats and mucks, and inorganic that is mineral sediments. Generally, aquatic plants grow and prosper in fresh, brackish, and saltwater but are found most commonly in freshwater in abundance.

Their habitats include flowing waters like rivers, streams, brooks, standing or stagnant waters like lakes and ponds, and the wetlands like bogs, fens, marshes, and swamps, which are classified as riverine, lacustrine, and palustrine communities, respectively.

At times the wetlands plants are regarded as helophytes. Marshes are influenced, that is, covered more than half by herbaceous species and swamps by woody species. Bog plants are aquatics that grow in acidic organic soils, whereas Fen plants happen to be growing in alkaline organic soils. Aquatic plants, which are also known as hydrophytes, macrophytes, and water plants, take place throughout the plant kingdom.

'Macrophyte' is a term that discriminates them from the microscopic aquatic algae which are not regarded as true plants. The whole world of aquatic plants is said to be evolved constantly over and over, which contains more than 250 independent origins by evaluation.

They tend to occur in the plants that are spore producing, for instance, ferns, liverworts, lycopods, and mosses but are sparse among the non-flowering seed plants like the gymnosperms with bald cypress (Taxodium) a remarkable exception.

Figure 1.5: An illustration of taxodium.

Source: Image by Flickr.

Flowering plants known as the angiosperms holds the substantial hydrophyte variation with more than species in proportion in monocotyledons as compared to in dicotyledons. However, hardly 2% of flowering plant species are aquatic.

1.3.1. Life Forms

Nevertheless, of their taxonomic affinities, the classification of aquatic plants is done ecologically by analyzing their forms of life. Categories consists the following:

1. The floating plants (*acropleustophytes*), with their stems and leaves that entirely float on the surface of water and their stems are not rooted in the bottom, such as duckweed (*Lemna*) and water hyacinths (*Eichhornia*) as discussed earlier.

2. Emergent (*hyperhydrates*), with stems and leaves extending mainly above the water surface and stems rooted in the bottom, such as cattails (*Typha*) and reeds.
3. Phragmites (planmergents or ephydrates), floating-leaved, with some or all leaves floating on the water surface and stems rooted in the bottom, such as floating-leaved pondweed (*Potamogeton natans*), water chestnut (*Trapa natans*), and water lily (*Nymphaea*).
4. Submersed (*hyphydrates*), with stems and leaves entirely underwater and stems embedded in the bottom, for example, Eurasian watermilfoil (*Myriophyllum spicatum*) and wild celery (*Vallisneria*).
5. Hanged (*mesopleustophytes*), with stems and leaves completely underwater and stems not rooted in the bottom, for instance, the bladderwort (*Utricularia vulgaris*) and the coontail (*Ceratophyllum*).

The use of Benthophyte and pleustophyte are done in order to distinguish the forms that are either rooted in the underlying substance or are unrooted. Species having extended leafy stems are labelled as vittate or caulescent like the coontail.

The leaves huddled together for the plants in a basal rosette are rosulate like the wild celery and those are not specifically recognized into stems and leaves are thalloid such as the duckweed. The differences in the floating and emergent leaves of the species morphologically from their submersed leaves are known as heterophyllous, for instance, floating-leaved pondweed.

1.3.2. Adaptations

Water plants are lessened anatomically and structurally. The angiosperm which is the smallest in the world, carries in itself plants that are only about 0.4 millimeter long. Submersed species have insufficient water at times conducting tissue (xylem), mechanical tissue (sclerenchyma), and cuticle.

On the other hand, there are some of them which don't have roots, they completely lack them. Support and floatation of submerged stems are lodged by buoyant tissue (aerenchyma) and substantial air spaces (lacunae) which helps in transporting oxygen all over the plant.

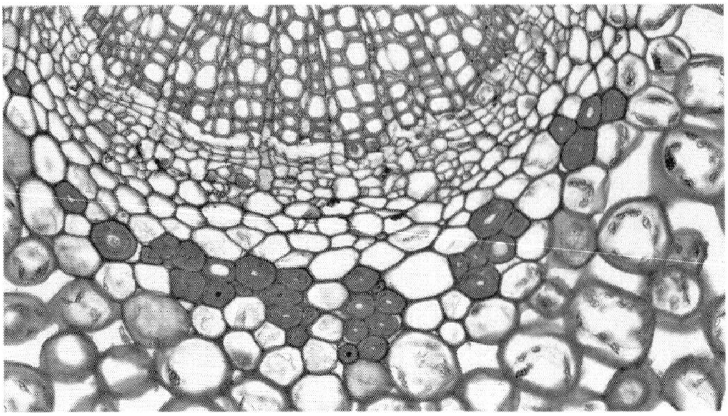

Figure 1.6: An illustration of midrib sclerenchyma.

Source: Image by Flickr.

Submersed plants normally contain either highly anatomized compound or thin, ribbon structured leaves. Few of the leaves become fenestrate, that is they do not have tissue between the veins. Such leaf shapes expand the surface area-to-volume ratios for more effective nutrient uptake and to deduct damage from water currents.

The leaves that float is generally flat and round and contains stomata on the upside surfaces. They often reach 2.5 meters in diameter like Victoria. For firmness, the stalks, i.e., petioles of most of the floating leaves are located in the center by margination of the base, as in the case of water lily, or are shield-shaped by the entire fusion of leaf lobes, as in the water shield that is Brasenia and Victoria. The aquatic plants deal with the harmful effects of anaerobic hydric souls through their physiological adaptations.

1.3.3. Reproduction

Most of the aquatic plants are incessant that reproduce vegetatively. That is they reproduce asexually. Species tolerate and live throughout winters or any other periods that have adverse effects as intact plants, by dying back to stem apices, by means of stems that are modified (rhizomes, stolon's, tubers), or other way doing is by making use of expertise dormant structures (hibernacula) in the sediment.

Winter buds are a kind of hibernaculum. These buds are enveloped by normal foliage leaves on abbreviated internodes. Most of the time, they remain connected to the plant. The specialized form of hibernacula are turions that produce modified, morphologically recognizable leaves to safeguard the enclosed buds. Turions have the ability of detaching itself from the plant and operates as propagules for dispersion. The fragmentation of stems causes the vegetative dispersion in water plants. That have brittle feature, because of the scarcity of mechanical tissue.

Detached stems can generate themselves by the production of adventitious roots. Several aquatic plants that are annuals produce seeds as their dormant stage, among which some of the aquatic annuals also grow and accumulate vegetatively by segmentation during the course of the growing season. Mainly the process of sexual reproduction occurs infrequently in submersed species, it is more prevalent in floating-leaved species, and fairly common in emergent species (and annuals).

Pollination in water plants is facilitated by insects that is entomophily, wind that is anemophily, and water or hydrophily. The pollinations through various sources results in most aquatics to be insect-pollinated and approximately one-third of them are wind-pollinated.

Not more than 5% of aquatic species are known as hydrophilous, with pollen transported on the water surface ephydrophily or under the water surface hyphydrophily. Most of the marine angiosperms, i.e., seagrasses are hydrophilous.

Water currents and waterfowl are responsible for scattering seeds, fruits, and the vegetative propagules. Although waterfowl are responsible for dispersing them on a large scale as compared to water currents.

They transport propagules in plumage, in mud adhering to their feet, and by the discharge of seeds consumed as food. A large variety of water plants are dispensed widely where some of the species get the worldwide dissemination.

1.4. TYPES OF AQUATIC PLANTS

In a pond, there can be various types of aquatic plants. It is not necessary that they are only one of a kind, and special care should be taken to recognize all the aquatic plants that are present and occupy the pond. Among them, some of the plants present in the pond might be useful to local or migratory wildlife, and hence it may want to be vitalized instead of removing them.

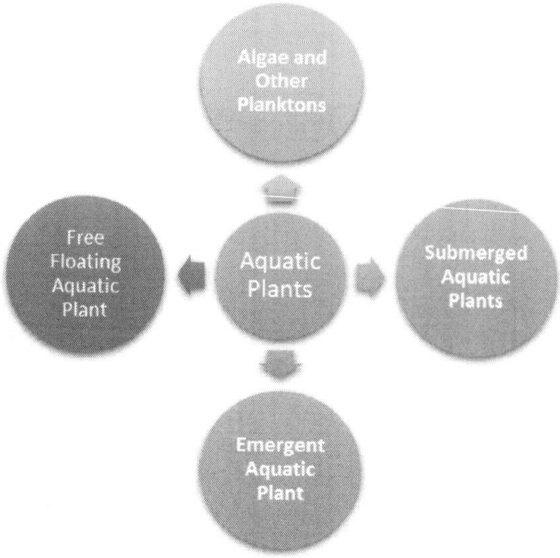

Figure 1.7: Types of aquatic plants.

As the temperature of water increases through spring and during the month of summer, then during this period aquatic plants starts growing. They are also important to the inhabitants in the ponds, although the concern is of way too much growth and of some non-native aquatic plants can result in a serious problem.

It is beneficial to comprehend and identify the types of plants that are emerging in the water bodies. It can be helpful in determining the management skills. So, generally, the aquatic plants are divided into four types, namely: Submerged, Free-floating, Emergent, Algae, and other Planktons.

1.4.1. Submerged Aquatic Plants

The primary quality of submerged plants is embedded in the bottom of the plant, and they start to grow up throughout the water column. They are found in water that is less than ten feet deep, yet some species are capable of growing at the depths of 20 feet.

Young fishes such as baby bluegill and other smaller species of fish like minnows make use of stands of submerged plants as a shield to protect and hide themselves from their predators. Furthermore, the submerged plants assist in diffusing oxygen to the water which eventually plays a significant role in maintaining a clean and healthy pond.

Whereas the growth of some submerged plant proves to be a good thing, if the growth of the plant becomes way too thick fish might avoid getting into that specific area. The flow of water becomes limited, and the oxygen level which was diffused gets dropped.

Figure 1.8: An illustration of submerged water plants.

Source: Image by Pixabay.

This results into anoxic conditions. In the United States, some plants are submerged and comes under this category. It includes eelgrass, elodea and American pondweed. The non-native species consists of hydrilla and parrotfeather milfoil.

These non-natives have the ability to grow very quickly and if they are not looked after, they will fill a pond entirely from top to bottom. In short, submerged plants are rooted plants having flaccid and rooted stems, and a large part of its vegetative mass is below the surface of water through small proportions of it might stick above the water surface.

1.4.2. Emergent Aquatic Plant

As submerged aquatic plants, emergent plants are rooted in the bottom of the plants, yet these species expand above the water surface. The stems of emergent plants are somewhat stiff and firm. This kind of emergent includes the floating emergent that is it refers to the plants that float on the surface of the water and are embedded in the bottom of the pond and put straight emergent that expands throughout the water column and draws straight out of the water.

Most commonly, these plants are to be discovered in wetlands and also along the shoreline, that grows in water generally for about four or five feet deep. The best examples for emergent erect aquatic plants are Bulrushes, and cattails whereas water lilies are the common floating emergent.

A lot many emergent plants are held as an aesthetic addition to the pond, yet if way too many of the plants are added in the pond, it may cause trouble with examining the pond by foot or by boat.

1.4.3. Free Floating Aquatic Plant

The main characteristic feature of the free-floating plants is that they are not rooted deep down in the soil. These plants live on the surface of the water. Although not all of them yet few of them consists of root that are suspended in the water from the floating green portions.

They are most commonly found in areas where there is a very small amount of movement or no water movement at all. As stated in the beginning of the chapter, the most prevailing free-floating plant species are Duckweed and water meal whereas the fast-growing non-natives are water hyacinth and giant Salvinia. These plants prove to be a good habitat for insects that in exchange supplies food for the fishes who live in the pond. Moreover, these are also habitat for unwanted pests such as mosquitoes. These plants are not harbored in the soil, the best and easiest way to take control over them is to simply move the water. There are various methods applied for the water movement and circulation, including diffused aerators and circulators. The vegetation that is most required by the pond is the aquatic vegetation as it adds beauty and representation to the pond. It also supports in keeping the ecosystem of the pond balanced although with a proper management. The larger growth of native species and the existence of non-native species may result in the decline of fish habitat. Secondly, it can also lead to harmful and anoxic conditions. In this case to the access with boats and foot is restricted.

1.4.4. Algae and Other Planktons

Algae are ancient plants. Some of them being microscopic such as the planktonic algae, while the others are thin and stringy, or they are hair like such as the filamentous algae. On the other hand, some of them are large and looks like higher plants but in the absence of true roots (chara).

Shifting the focus towards algae, they are plant like organisms. They possess simple different structures from single-celled phytoplankton that float in water to a large number of seaweeds that is macroalgae that are connected to the ocean bed. The most common places where algae are to be found are oceans, lakes, rivers, ponds, or they can be even found in snow, be it any place on earth.

Algae are often called primitive plants, yet they are plant-like and not regarded as plants. For algae, protists are the most suitable term, mainly for a single-celled phytoplankton. Although, more complex algae like kelp and chara are at times mistaken for submerged plants.

Seaweeds and submerged plants differ in their structure. The feature of macroalgae is that they are simpler and connect themselves to the seabed not with the true roots but with a holdfast whereas the other aquatic plants like the floating, emergent and submerged have distinguished parts such as the roots, stems and leaves. Most of the plants contain vascular structures, i.e., xylem and phloem.

Figure 1.9: An illustration of algae in water.

Source: Image by Flickr.

This helps in carrying nutrients throughout the plant. On the other side, algae have chlorophyll just like plants do, and they lack the specialized structures. It's difficult to classify algae as they can be single celled, filamentous or plant like.

Most of the time, algae are grouped by their primary color, such as green, brown, and red which prove to be a problem instead of solving it. The different species of algae widely vary from one another. They vary in pigmentation, cellular structure, complexity, and chosen environment.

Due to these various reasons, algal taxonomy remains a matter of discussion. Some classify algae under different kingdoms including, Plantae, protozoa, and chromista. The classification of overarching kingdom is not usually accepted and agreed on the basis of the species, genus, family, class, and phylum of each algae.

1.4.4.1. Phytoplankton

Phytoplankton are referred to as microorganisms that float freely in water bodies. Usually, they are single-celled, but sometimes they expand themselves in colonies large enough that are visible to a human eye. The nature of phytoplankton is photosynthetic, meaning to say, they are capable of making use of the sunlight in order to convert carbon dioxide and water into energy.

Although they possess the ability of plants yet they are not regarded completely as plants. When people use the term misnomer, it is regarded as inaccurate, and it should not be used. The classification of phytoplankton is done in two classes.

They are – algae and cyanobacteria, respectively. These two classes include the common capability of doing the process of photosynthesis, yet they contain physical structures that are different. Nonetheless, whatever taxonomy they have, all phytoplankton carry at least one form of chlorophyll mostly chlorophyll A. Using this they can carry on the process of photosynthesis for gaining more and more energy.

Both the classes of phytoplankton be either algae or cyanobacteria, can be discovered in fresh or saltwater. As there is a requirement of light to go on with photosynthesis. No matter what the environment is phytoplankton floats close to the top of water where the sunlight reaches.

Most of the phytoplankton are built up of green algae and cyanobacteria, which is usually known as the blue-green algae. Talking about the marine phytoplankton, they mainly consist of microalgae which is called the dinoflagellates and diatoms, although few other algae and cyanobacteria can be available and present.

Some autonomous movements are seen in dinoflagellates because of their tail which is flagella, but diatoms are entirely dependent upon the currents of the ocean.

1.5. USES OF AQUATIC PLANTS

There are various benefits of aquatic plants which help and plays a vital role in promoting the diversity and function of aquatic ecosystems. A large amount of production in the aquatic system takes place either in marine or littoral areas that is the freshwater are, this is a function of both macrophytes and algae.

These productive areas have the main characteristic of being shallow, have relatively firm chemistry in water and sufficient amount of light reaches to the bottom surface in order to back up the growth of aquatic organisms, as stated by Dodd in 2002.

In the case of the freshwater plant species which is most submersed, the growth at the depth of water whereas estimated the percentage of light that reaches to the bottom substrate is 21%. This statement was given by Chambers and Kalff in 1985.

The huge amount of light present and its greater availability and also the warmer water temperatures, phytoplankton, zooplankton and other aquatic organisms flourish in these areas and regarding marine habitats, hold up some of the best fisheries which are eminent in the world.

In the aquatic ecosystem, algae shape the base of aquatic food chains. According to Smith, Algae are the primary producers, they are in charge of producing more than 70% of the world's oxygen (2011). There are many different growth forms of algae that can exist as either single cells or complex multi-cellular forms, for instance, filamentous, sheets, or cylindrical forms.

Generally, the algae that are small, single-celled are often knows as phytoplankton, whereas the multicellular species that are larger are named and described as macrophytes, if growing in freshwater or seaweeds that is macro-algae, if growing in the marine environments. (Smith, 2011).

1.5.1. Marine Habitats

The role of macroalgae is diverse; it serves many functions inside the marine environment, including serving as the foundation of the food chain for humans as well as animals. The important categories of algae include Laminaria, Macrocystis, Nereocystis, Palmaris, Ulva, Undaria, Fucus, Porphyria and Saccharina.

For other animals and small zooplankton, macroalgae plays an important role as refuges. It also performs its duties as a direct food source for various animals that are huge along with marine amphipods as stated by Cruz-Rivera and Hay, 2000.

Figure 1.10: An illustration of laminaria.

Source: Image by Flickr.

Moreover, macroalgae form large floating wracks as they get displaced. These floating wracks cater the ecological functions continually. "They also prove to be significant sources of recycled nutrients in near-shore seagrass meadows and reefs, and the detritus produced as a result of these wracks from the basis of the food chain" (Robertson and Hansen, 1982; Kirkman and Kendrick, 1997). The decomposition of algae in the coastal areas certainly are an essential source of nutrients for subtidal communities (Robertson and Hansen, 1982; Walker et al., 1988). The main source of particulate carbon are the surf zone and beach cast algae wracks.

These carbon supports the food webs that are the nearshore detrital based which contains suspension feeders close to the shore fishes and also beach waders. In places like California, when the amounts of kelp wrack that wash ashore on beaches increase, the black and ruddy turnstone populations increase as contented by Bradley and Bradley (1993).

Wrack inhabiting organisms are responsible for feeding bird species. For instance, crustaceans, molluscs, insects, and polychaetes that make use of algae wracks (Kirkman and Kendric, 1997).

1.5.2. Freshwater Habitats

According to Carpenter and Lodge, the significant role played by the plants in the littoral zone are wide ranging. This extensive importance is possible because they contribute to the structure, function, and diversity of aquatic ecosystems, aid in a cycle of nutrients, becomes a source of food for aquatic animals along with providing habitat for invertebrates and fish (1986).

Aquatic plants are responsible for anchoring soft sediments, stabilizing the slopes underwater, wipe out nutrients from the overlying waters. The insufficient amounts of submerged plants are in charge of resuspension of sediments that are at bottom that occurs often and environments that have low lights, both these things have an adverse effect on the growth of submerged plants (Chambers and Kalff, 1985, Barco et al., 1986, Schiffer, 1998).

Young and adult fishes of many species are dependent upon aquatic plants during their life course, and then eventually they move to different habitats depending upon their growth stage. Numerous fishes among them rely on habitats that have aquatic vegetation, for about 120 species, representing 19 families of fishes prove to have a larger quantity of fish in comparison to the non-vegetated areas.

The aquatic vegetation acts as a protective shield for young fishes in order to hide from predators, and they are also dependent upon algae and the microfauna, for instance, zooplankton, insects, larvae for their food, that live on aquatic plants (Dibble, 2009). In general, nesting, growth, foraging success of numerous species of fish are demonstrated by the composition and density of plants.

1.5.3. Other Uses of Aquatic Plants

Both the varieties of aquatic plants that is marine and freshwater are used widely as livestock fodder, fertilizer, compost, mulch, and bioremediation.

Many of the aquatic plants consist of a lot of crude protein, crude fat, and the mineral matter as many compact forage crops on a cornerstone of dry weight. The fiber values were higher in forages and lower in aquatic plants.

The utilization of aquatic plants as a fodder will assist in paying for harvesting, which is known as the proper and accurate way of separating nutrients from the lakes that suffer from artificial enrichment. The best time for doing harvesting is when the content of protein of the plants is comparatively most high for their utmost usefulness as fodder.

However, Hasan and Chakrabarti in 2009 gave reviews on the uses of aquatic plants as feed-in aquaculture production, globally. They stated that whatever the recent circumstances are algae may not be a feasible choice as a feed for aquaculture production. Although an analysis of the cost-benefit is required before reaching to any point and making a concrete interpretation for its use as fish feed.

The species of duckweed like Lemna, Wolffian, spinodal, and Landolina was proved to be important as they gave an entire feed package for polyculture, yet in a few countries, the year-round availability of duckweed might be troublesome for its extensive and global adoption.

Figure 1.11: An illustration of lemna (also known as common duckweed).

Source: Image by Flickr.

1.5.4. Economic Uses of Aquatic Plants

Over the past few years, the demand for aquatic plants, especially the marine macro-algae have expanded rapidly. The overall harvest of aquatic plants in 2009 were estimated at around seventeen million tons, in which the marine algae comprised 85% of the aggregate.

Countries that contribute to the large number of wild-harvested macroalgae is China, Chile, Norway, Japan, and Russia. The production of macroalgae as a result of aquaculture is much more in China than in Indonesia, the Philippines, Korea, Japan, Malaysia, and Viet Nam according to a report by FAQ (2010).

The freshwater plants harvest has been slightest. It ranges from no reported harvest from the year 1950 to 1966, towards the harvest of around two million tons done every year. The increase in demand in food, pharmaceutical, biomass feedstock has influenced and increased in the harvest of marine aquatic plants on the other hand, the increase in the harvest of freshwater plants is due to the popularity of water gardening and aquarium plantings.

Among which many of the species belonging to water garden are imported from tropical and subtropical regions. As stated by a report, since the year 1990, the global value of aquatic plants has been between US, 4.8 billion dollars, with the highest increase in the year 2007.

The harvest statistics depict that marine macroalgae are influencing the value of aquatic plants up to a great extent all over the globe. Freshwater aquatic plants consist of a small amount of a total value from the eyes of looking globally.

In United States, water gardening is a large industry for about US$1 as determined by the retail sales. Talking about Europe, the sales volume of aquatic plants there were estimated to be seven million plants in 2006 and 2007, respectively, and approximately two million plants were sold by a Danish company and in addition to this more 2.1 million plants were traded from the countries like Near East and Asia, according to the data given by Brunel in 2009.

As the globalization of commerce, trade and industry expanded through the internet and enhancement of shipping methods, an extended number of sales are made with species transported to larger distances at way too increased rates.

Dissimilar to the harvest statistics, where kelp and red seaweeds were composed of the greatest volume, Caulerpa seaweeds since the ancient time has been considered of highest value and importance in all the other species. The value of Caulerpa seaweeds hanged between US$2 and 3 billion from the beginning of the 1990s until 2007 when a gradual decline took place.

In conclusion to this, aquatic plants have larger than life benefits, their uses are reaching to people across the globe day by day, yet their new uses are still remaining to be discovered. In the recent times, due to the introduction of aquatic plant species, both marine and freshwater habitats are being intimidated, which proves to be troublesome under certain situations.

These species of plant are most of the times are instituted from other parts of the world for various reasons such as the beneficial or horticulture uses and thereby escape cultivation in order to form natural populations.

According to Shea and Chesson, aquatic plants are at a frequent risk of colonization by plant species that create problem due to the constant disturbance that acts as a catalyst for the growth of these species (2002). The transformation of biotic and abiotic interactions usually occurs when the troublesome plants occupy an area (Madsen, 1998).

The growth of problem species at times has an outcome which is depletion in more desirable plant species, reduced production of fish (Savino and Stein, 1989) and enlarged sediment resuspension, turbidity and algal production.

1.6. AQUATIC PLANTS AND ECOSYSTEM

Aquatic plants possess a wide spread range of ecological roles and make a considerable contribution to the structure, function, and service provision of aquatic ecosystems. As mentioned, their well prevailing significance in aquatic ecosystems, research into aquatic plants continues to flourish.

The 14th International Symposium on Aquatic Plants, which was held in Edinburgh in September 2015, conducted together cooperatively where 120 representatives from 28 countries and six continents were assembled together.

This peculiar issue of Hydrobiology consists of a select number of papers on dimensions of aquatic plants, which covered a wide range of species, systems and issues. The conference dealt with the understanding of aquatic plant biology, the aspect of scientific issues being observed and addressed and the range of techniques available to researchers have all arguably never been greater; although substantial challenges exist to the preservation and management of both aquatic plants and the ecosystems in which they are found.

The range of countries and continents represented by conference delegates and authors of papers, different research, and studies in the special issue demonstrates the global importance of aquatic plant research in the early 21st century, but also it throws light on the many challenges on this burgeoning scientific discipline.

The key concern about the biogeographic observation on aquatic plants is their frequency of endemism is less than that of terrestrial plant comparatively. Most of the species of aquatic plants are extensive and spread all over the world. They exist on more than one continent due to the far-reaching dispersal by natural vectors mainly and recently by human beings. However, each species has its own level of tolerance and adaptation when it comes to the environment.

Harsh environmental conditions for aquatic plants are characterized by some mixture of dryness, a flow velocity on the surface and high salinity, along with the persistent ice cover that is short growing season for about nine months and the growing season temperature at high latitude or altitude in thermal habitats.

Aquatic plants are capable of dealing and tolerating with wide dimensions of environmental conditions while few of them have a low tolerance level and becomes important as environmental indicators. The most common example of environmental indicator is macrophytes which are tactful to particular thresholds of environmental conditions, and eventually, they are important indicators for change.

Although species that are sensitive to one environmental aspect are not reactive in the same way to the other. So, it is important to know the tolerance level of species dependent upon various environmental factors then only aquatic plants are used as indicators enclosed in an ecoregional context. The aspects that influence the most is the complexes of factors such as chemical, physical, and biological and how they effect on individual aquatic species and communities.

More likely, most of the species of aquatic plants depict a high level of polymorphism and phenotypic plasticity as they respond and react to it. This is shown in their responses to differences of environmental factors, a feature that permits them to occur over a larger dimension of conditions.

Due to this inherent biological difference, most of the macrophytes prove to be poor indicators of particular habitat conditions. In some cases, the species have narrower tolerance, yet they are very vital indicators of environmental conditions. This can be in two ways, either in the context of their presence and availability or relative abundance within the communities.

1.7. CONCLUSION

When it comes to plants, everyone knows that they are of different kinds. People have been made familiar with plants that are on land, but what about the ones underneath the water? They are the category of plants that are meant to live underwater. Aquatic plants are very fascinating and have unique characteristics, and this chapter deals with the same uniqueness, features, forms, and types of aquatic plants.

Any type of plant or vegetation which lives and grows in water is called an aquatic plant. They are found in all types of water from seawater to freshwater. For instance, they are found and visible in oceans, rivers, lakes, ponds, and more.

Just like aquatic animals that have unique characteristics, aquatic plants also possess features that the terrestrial plants do not possess. They are special in their own way. However, they provide oxygen to the animal species that live under the ocean bed. It is very crucial for their metabolic processes. Moreover, they also supply food to some of the species of animals living underwater. For example, the turtles eat the algae present in freshwater pond surfaces.

While some remain underwater, some of the aquatic plants float above the surface of freshwater. Their stems and roots allow them to remain firmly affixed despite strong currents. For instance, moss clings on to rocks. The various forms of aquatic plants have been discussed in the chapter. Tracing from the history to the life process, its different uses and their survival and affect in the ecosystem, this chapter sheds light on every aspect.

REFERENCES

1. AquaPlant. (n.d.). Plant Identification | *Pond Algae, Grass, Weeds, Moss, and More*. [online] Available at: https://aquaplant.tamu.edu/plant-identification/ (accessed on 31 July 2021).
2. Dhir, B., (2021). Physiological basis of arsenic accumulation in aquatic plants. *Handbook of Bioremediation,* [online] pp. 237–244. Available at: https://www.sciencedirect.com/science/article/pii/B9780128193822000144 (accessed on 31 July 2021).
3. Encyclopedia, (2018). Aquatic Plants | *Encyclopedia.com*. [online] Available at: https://www.encyclopedia.com/plants-and-animals/botany/botany-general/aquatic-plants (accessed on 31 July 2021).
4. Fondriest Environmental, Inc., (2014). Algae, Phytoplankton and Chlorophyll-Environmental Measurement Systems. [online] *Environmental Measurement Systems*. Available at: https://www.fondriest.com/environmental-measurements/parameters/water-quality/algae-phytoplankton-chlorophyll/#:~:text=Algae%20are%20sometimes%20considered%20protists,are%20often%20difficult%20to%20classify. (accessed on 31 July 2021).
5. Kasco Marine, (n.d.). Three Types of Aquatic Plants to Watch for This Spring | *Kasco Marine*. [online]. Available at: https://kascomarine.com/blog/3-types-plants-watch-spring/ (accessed on 31 July 2021).
6. Lacoul, P., & Freedman, B., (2006). Environmental influences on aquatic plants in freshwater ecosystems. *Environmental Reviews,* [online] *14*(2), 89–136. Available at: https://www.researchgate.net/publication/233499337_Environmental_influences_on_aquatic_plants_in_freshwater_ecosystems#:~:text=Aquatic%20plants%20are%20important%20components%20of%20many%20freshwater%20ecosystems.&text=Some%20aquatic%20plants%20have%20a,a%20wide%20range%20of%20conditions. (accessed on 31 July 2021).
7. O'Hare, M., Aguiar, F., Asaeda, T., Bakker, E., Chambers, P., Clayton, J., Elger, A., Ferreira, T., Gross, E., Gunn, I., Gurnell, A., Hellsten, S., Hofstra, D., Li, W., Mohr, S., Puijalon, S., Szoszkiewicz, K., Willby, N., & Wood, K., (2017). Plants in aquatic ecosystems: current trends and future directions. *Hydrobiologia*, [online] *812*(1), pp. 1–11. Available at: https://link.springer.com/article/10.1007/s10750–017–3190–7 (accessed on 31 July 2021).

8. Plant Life, (n.d.). *Aquatic Plants*. [online] Lifeofplant.blogspot.com. Available at: http://lifeofplant.blogspot.com/2011/12/aquatic-plants.html (accessed on 31 July 2021).
9. The Aquatic Plants of Mountain Lakes. (2017). [ebook] Hackettstown, NJ: *Borough of Mountain Lakes,* pp. 1–18. Available at: http://mtnlakes.org/wp-content/uploads/2014/09/MtLakesAqPlGuide2017.pdf (accessed on 31 July 2021).
10. Toppr, (n.d.). *Aquatic Plants – Definition, Types, and Importance*. [online] toppr.com. Available at: https://www.toppr.com/guides/biology/plants-and-mushrooms/aquatic-plants/ (accessed on 31 July 2021).

Chapter 2

Biology of Aquatic Plants

CONTENTS

The current chapter provides an insight on the biology of aquatic plants. It discusses the anatomy and anatomical features, namely, the root, stem, leaves, and flowers of aquatic plants as also the morphological adaptations of aquatic plants. The classification of aquatic plants into its main life forms, i.e., floating, submergent, and emergent aquatic plants has been discussed in detail.

The chapter further looks into the evolution of aquatic plants and also their distribution and diversity across the globe. The physiology and cellular biology with special emphasis on photosynthesis and cellular structure have been discussed towards the end of the chapter.

2.1. INTRODUCTION

Certain plants in aquaculture are classified as aquatic plants as they are plants that are accustomed to living in aquatic environments such as saltwater or freshwater. They are also referred to as hydrophytes or macrophytes to differentiate them from algae and other microphytes.

Macrophytes are classified as plants that either grow in or near water, and they can be emergent, submergent, or floating on the water. In lakes and rivers, macrophytes provide cover for fish, substrate for aquatic invertebrates, produce oxygen, and act as food for some fish and wildlife.

Figure 2.1: Aquatic macrophytes.

Source: Image by Flickr.

Aquatic plants are plants whose photosynthetically active parts are perpetually or at least, for several months each year submerged in water or float on the surface of water. All vascular plants that may be encountering growing in permanent or seasonally semi-permanent water are included.

Several plants, particularly those that thrive in the tropical regions and those which grow on rocks or trees by streams, rivers, and waterfalls either are submerged in swiftly flowing water, sometimes for only a few hours or days at a time during the rains. These plants are called rheophytes and although specialized for this kind of habitat.

Freshwater macrophytes play a significant role in the aquatic ecosystems. They are responsible for supplying, either directly or indirectly, food, shelter, and a variety of habitats for a large number of organisms, including wildfowl and economically important fish. Rice, an aquatic plant, is the most important single crop species in the world.

Many other aquatic plants are also of use to mankind as food, raw materials for industrial processes, building materials and manure in agriculture. Aquatic plants absorb dissolved minerals and enrich water with oxygen produced during photosynthesis. These properties are of benefit to man as they assist in the maintenance of clean water and the help in the recovery of polluted water.

However, in disturbs or newly constructed bodies of water, rampant growth of aquatic plants may interfere with man's use of fresh water. They may obstruct water flow, navigation or water intake. They may also create conditions favorable for pests, diseases, and vector affecting humans, animals, and crop plants.

2.2. AQUATIC PLANTS ANATOMY

Plant anatomy is the study of the shape, size, and structure of a plant. Although certain aquatic plants lack a central stem, and plants such as mosses and ferns do not produce flowers altogether, the anatomy of most plants can be divided and studied under four basic areas; the root stem, leaves, and flowers. All of these parts play a pivotal role in the plant's basic functions, including growth, reproduction, nutrient-collection and storage.

Aquatic plants, like all other typical plant bodies, consists of three major vegetative organs, namely the root, the stem, and the leaf, as well as a set of reproductive parts that comprise flowers, fruits, and seeds.

2.2.1. Root

Most aquatic plants have roots that are an amalgamation of numerous central roots, with many smaller roots trailing off. In a sharp contrast to terrestrial plant roots that have profusely developed fine hairs for trapping moisture, in aquatic plants, there is a complete absence of root hairs, although they may develop when grown out of water. Some aquatic plants produce roots from rhizome that attaches to rocks and wood.

Submerged plants, unlike terrestrial plants, lack the external protective tissues required to reduce water loss. As a consequence of xylem atrophy in aquatic plants, the internal system of tubes (xylem) which normally transports water from the roots to all parts of the plant is often poorly developed, if not absent.

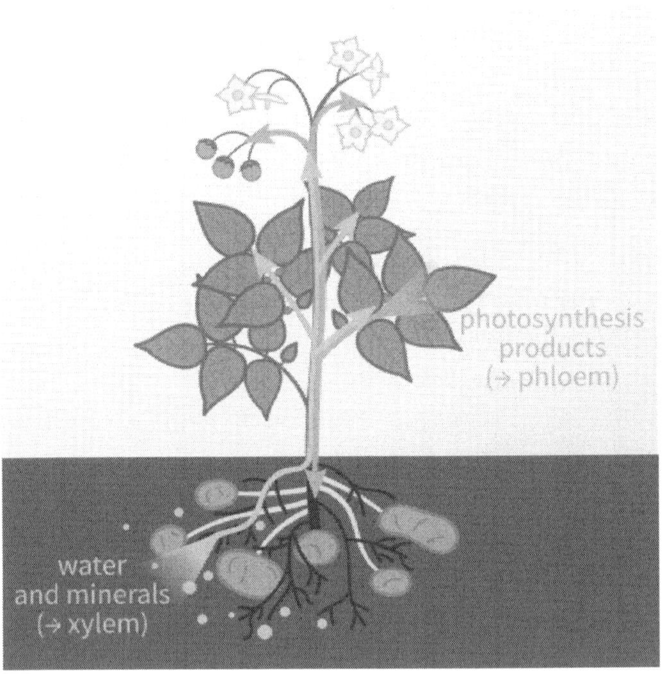

Figure 2.2: An illustration of the role of xylem in aquatic plants.

Source: Image by Wikimedia Commons.

The primary function of roots in aquatic plants, as in the case of land plants is to provide anchorage and the absorption of water along with minerals and nutrients. Floating plants are provided buoyancy by roots, whereas submerged plants receive anchorage. Since aquatic plants are surrounded by water, water containing absorbed nutrients is absorbed all over the plant surface. Plenty of aquatic plants grow from bulbs or tubers, which act as large reservoirs of nutrients.

Examples of aquatic plants that have deep, anchoring root systems that keep them affixed to the soil at the bottom of the water body are seaweeds, eelgrass, and pondweeds, whereas some floating aquatic plants, like duckweeds float freely and allow their roots to drift in the water below them. One aquatic plant is eelgrass, which has adaptation for living in saltwater.

2.2.2. Stem

A stem is present in most aquatic plants and performs two basic functions: support and transport. The stem's function is aided by supporting gas- or air-filled cell that provide buoyancy and help to keep the plant upright. Since the surrounding water provides much of a plant's support, aquatic stem is often much thinner more flexible than terrestrial stems. Terrestrial plants have hard stems to conserve water and stand straight to gain sunlight.

While aquatic plants have soft and light stems due to the buoyant force of water, they need not much rigidity to rise up for gaining sunlight. The stems of many aquatic plants have large intercellular air-filled spaces known as air vacuoles to increase buoyancy and help free-floating plants remain lightweight. Flexible stems allow the plant to move with the water, rather than try to hold steady against it, risking damage.

2.2.3. Leaves

The leaves of a plant, collectively referred to as foliage are essentially tools for collecting sunlight to use in the process of photosynthesis. Leaves perform main functions such as producing food and interchange of gases between the water body and the plant body and collection of nutrients is also carried out by the leaves.

The leaves of terrestrial plants have a thick, waxy outer layer called cuticle, which protects the plant from drying out. In aquatic plants, this layer is much thinner and liquid able to pass through much more easily, which help to plant to take up nutrients.

Aquatic Plants that live over the water have broad leaves in order to maximize the loss of water through evaporation because they live in an environment where water is in excess. Moreover, the large surface area helps them in photosynthesis by absorbing a greater amount of light.

Figure 2.3: An illustration of leaves of duckweed plants.

Source: Image by SnappyGoat.

Plants that live under the water in the sea have wax on their leaves. Because water is salty in the sea and due to hypotonic environment outside the leaves, water tends to lose through their leaves. Therefore, they have protection of a wax coating to save themselves to withstand the decaying effect of water and eventual rotting.

Aquatic plants that produce aerial leaves often display two different leaf shapes below and above the water. This is due to different environments and a change of cuticle layer. Plants that grow underwater will usually have leaves that are thinner and softer, while the leaves that grow in the air will be much thicker and have a harder surface. They often have thicker stems too.

Shining, glossy surfaces are also known to slow down transpiration to some extent. It reflects back falling sunlight and reduces the heating effect on leaves. Hydrophytes have poorly developed water-conducting system as they are able to absorb water through their general body surface. Thus, the shining surface of the leaves also cuts down transpiration and eventual water loss.

The foliage can have a totally different shape and form, too, for example, *Hygro difformis*, which appears like a finely divided fern when it's growing underwater, but has single, serrated leaves when it's growing emerged, in air.

2.2.4. Flowers

Although not all aquatic plants are likely to produce flowers in the aquarium, the majority are flowering plants and will produce seeds and reproduce by flowering in nature. The flowers are usually produced above water, where they can be pollinated by insects, just as terrestrials ones are.

A few examples of aquatic flowers are lotus, water lily, water iris, water Hawthorne, water poppy.

Some aquatic plants produce flowers beneath the water surface. In these instances, the seeds are capable of floating downstream, and a few species do not produce flowers at all, preferring to reproduce by purely asexual means.

2.3. ANATOMICAL ADAPTATIONS OF AQUATIC PLANTS

1. The root and shoot systems show common features such as cuticle which is very thin or absent.
2. Epidermis is usually a single layer of thin-walled cells, not protective in function.
3. Cortex is well developed. It has numerous air chambers. It helps in buoyancy and rapid gaseous exchange.
4. Mechanical tissues are generally absent.
5. 1In the vascular tissue, xylem vessels are less common. Only tracheids are present in submerged forms.
6. In amphibious form, the xylem and phloem are well developed (e.g., *Limnophylla heterophylla*) or vascular bundles may be aggregated towards the centre. (e.g., *Jussiaea*)
7. Epidermal cells of Phloem leaves contain chloroplasts and they can function as photosynthetic tissue, especially where the leaves and stems are very thin, for example, *Hydrilla*.
8. Stomata are totally absent in submerged, but in floating leaves, stomata are confined only to the upper surface. In amphibious plants, stomata may be scattered on all the aerial parts.
9. In submerged leaves, air chambers are filled with respiratory and other gases and moisture.

10. In Water Lilly (*Nymphaea*) and some other plants, special type of star shaped lignified cells called asterosclereids are developed. It gives mechanical support to the plants.
11. The aquatic plants exhibit a low compensation point and low osmotic concentration of cell sap.
12. Mucilage cells and mucilage canals secrete mucilage to protect the plant body from decay underwater.

Figure 2.4: An illustration of pink Nymphaea hybrid flower.

Source: Image by Wikimedia Commons.

2.4. LIFE FORMS OF AQUATIC PLANTS

Aquatic plants can be classified on the basis of their natural habits, ecology and growth form into the following categories.

2.4.1. Floating Plants

Floating aquatic plants are those plants that grow or live entirely on the surface of the water, with leaves, blooms, and even root systems floating in the water, one type of fixed plant closely resembles this growth pattern, as in Lemna and Riccia.

Another type of emergent plant, floating-leaf plants produce growth on the surface of the water and send long tendril-like root systems down to the soil bed at the bottom of the water body to affix the plant into place in the pond. Floating-leaf, affixed aquatic plants include such species as water lilies and water chestnuts, which have leaves and blooms that float above the surface of the water.

Free-floating macrophytes are aquatic plants that are found suspended on water surface with their root not attached to substrate, sediment, or bottom of the water body. They are easily blown by air and provide breeding ground for mosquitoes.

They obtain nutrients directly from the water through foliar absorption and through water roots. Examples of free-floating plants include Pistia spp, commonly called water lettuce, water cabbage or Nile cabbage.

Figure 2.5: An illustration of floating plant: *Potamogeton crispus*.

Source: Image by Wikimedia commons.

Floating-leaved macrophytes, on the other hand, are those that have root systems attached to the substrate or bottom of the body of water and with leaves that float on the water surface. Common floating leaved macrophytes are water lilies (family *Nymphaeaceae*), pondweeds (family *Potamogetonaceae*).

2.4.2. Submerged Plants

Submerged plants are aquatic plants which are rooted on the water's bottom, for example, Myriophyllum spicatum or without any root system, for example, *Ceratophyllum demersum* but do not extend all the way to the surface. These plants spend their entire life cycle beneath the surface of the water, with the possible exception of the stage of flowering.

The leaves, stems, and roots grow entirely and permanently underwater, although some of its leaves float. These types of plants tend to have very soft textures that allow small waving and flowing when underwater.

Many submerged plants include only leaves and stems and use root systems to spread. Submersed plants must obtain dissolved oxygen from the water as their leaves have the high surface area to volume ratio, and the cuticles are found to be absent. Examples of submerged plants include various species of seaweeds, eelgrass and pondweeds.

All photosynthetic parts submerged, are assumed to be the bottom rooted. In addition, they have flowers usually raised above the water surface. They usually grow near the shore up to the deepest part of the littoral zone. They are tolerant of fluctuating water levels, shoreline waves, and erosion.

Some common examples of submerged plants are pondweed, hornwort, and rice-field water-nymph.

Some species of submerged plants are known to have antipollution mechanisms. They are often used for phytoremediation – the use of plants to remove, degrade, and contain contaminants such as heavy metals.

Helophytes are plants that grow in a marsh, partly submerged in water, so that it regrows from buds below the water surface. Fringing stands of tall vegetation by water basins and rivers may include helophytes. Examples include stands of *Equisetum fluviatile*, *Glyceria maxima*, *Hippuris vulgaris*, *Sagittaria, Carex*, *Schoenoplectus, Sparganium, Acorus*, *yellow flag* (*Iris pseudacorus*), *Typha* and *Phragmites australis*.

2.4.3. Emergent Plants

Aquatic plants that generally grow in water but pierce the surface so that they are partially in air are classified as Emergent plants, and collectively they are called emergent vegetation. These plants have their roots and basal portions growing beneath the surface of shallow water, but their leaves and stems are primarily in the air and extended above the surface of the water. Examples of emergent plants include bulrush, cattails, arrowhead, rushes, sedges and a variety of shoreline plants.

Figure 2.6: An illustration of an emergent plant: bulrush.

Source: Image by Pxhere.

This peculiar feature may have developed because the leaves can undergo photosynthesis more efficiently in air and competition from submerged plants, but often, the main aerial feature is the flower and the related reproductive process. The emergent habit permits pollination by wind or by flying insects.

The leaves of these plants have spongy tissues and are packed with air spaces. The stomata on the leaves provide oxygen to emergent plants directly from the atmosphere. They typically grow along the shore where the water is low, usually less than four feet deep.

Some common examples of emergent plants are wild rice, cattails, pickerelweed, the reed (*Phragmites*), *Cyperus papyrus*, *Typha* species, and flowering rush species. Some species, such as purple loosestrife, may grow in water as emergent plants, but they are capable of flourishing in fens or simply in damp ground. A few species of emergent plants grow via stolons or rhizomes. Stolons, also called runners, are stems that allow the plants to spread and reproduce. These stems enable emergent plants to overcrowd other species and allow them to endure periods of environmental stress.

Contrarily, rhizomes are modified stems, where the shoots and roots arise through the nodes. The shoots and roots serve as the overwintering buds of the plants during cold weather.

Some photosynthetic parts in contact with air, assume to be bottom rooted with assimilating parts in air. They may also flower underwater or above water. This group includes many aquarium plants such as *Barclaya* and *Nymphaea*.

2.5. EVOLUTION, DISTRIBUTION, AND DIVERSITY OF AQUATIC PLANTS

Aquatic plants have unquestionably originated from terrestrial plants. The evolutionary journey of aquatic plants started and ended in the water. The first stage was the conquest of the aerial environment, resulting from the innovative capacity to achieve fertilization independently of water, and also from adaptations of vegetative organs to the dry environment.

This allowed dispersion of plant into the terrestrial environment, especially for the angiosperms, which is the most abundant group of plants today. This metamorphosis from aquatic to terrestrial environment required an evolution of some structures to permit water transport (roots and vascular tissues), aiming, not only, to minimize the water loss to the external environment, but also to maintain hydric flow (cuticle and stomata), and to provide support to organs and tissues (cellulose and lignin). Primitive plants irradiated to the terrestrial environment with the evolution of vascular plants and angiosperms, continuing the adaptive radiation that allowed the colonization of all environments distributed on the Earth surface. Eventually, a few species of bryophytes, pteridophytes, and angiosperms, especially herbaceous ones, returned to the aquatic environment, where they were able to survive and reproduce.

Not only the freshwater but also the saltwater environments were invaded by terrestrial plants. As the water excess limited growth and reproduction of these plants, the aquatic environment colonization by angiosperms resulted from the evolution of morphological, anatomical and ecophysiological characters.

Figure 2.7: An illustration of mousse bryophytes.

Source: Image by Wikimedia Commons.

It is estimated that the amount of plant species that are adapted to the aquatic environment is no more than 1% of the angiosperms and 2% of the pteridophytes. Among 250,000 species of angiosperms already discovered, only a small fraction of these is adapted to the aquatic environment, or even to soils saturated with water.

Until recently the group of angiosperms was divided in two classes—monocotyledons and dicotyledons. According to genetic evidence from several molecular biology research studies, this classification needs to be revised, with the creation of a third group, the magnoliid, the characteristics of which are more primitives.

Traditionally, this group was inserted in the dicotyledons, although there are some well distinguished characteristics. Inside the magnoliids, the following two groups are recognized:

1. woody magnoliids; and
2. paleoherbs magnoliids,

which is considered an isolated group according to the Actual Phylogenetic Group (APG II, 2003), widely used for angiosperms identification.

Contemporary botanists regard the paleoherbs to be the ancestors of the monocotyledons, appearing sometime before the end of the Cretaceous, more than 120 million years ago, and of the eudicotyledons, which are the dicotyledons with the exception of the magnoliids. Members of the Nymphaeales – Nymphaeaceae and Cabombaceae – and of the Ceratopyllales are the paleoherbs that occur today.

Many primitive families of monocotyledons are aquatic, indicating that the adaptive radiation to the aquatic environment occurred earlier in this group. Therefore, wetland and aquatic plants are more frequent among monocotyledons then eudicotyledons.

It is estimated that among all aquatic plants occurring in the planet, only 14% belong to the eudicotyledons, including submerged, free-floating, fixed with floating leaves and emergent plants, while 52% belong to the monocotyledon families. Certain authors have often suggested that the monocotyledons had their origin in the aquatic system, but this hypothesis is often challenged and not well accepted.

The use of DNA analysis to investigate phylogenetic correlation between aquatic and terrestrial plants may clarify and explain several questions that have still not been answered. DNA analyses of families that have terrestrial and aquatic species indicate that the terrestrial group is evolutionarily more recent.

Several aquatic plants retain in the aquatic environment some reproductive mechanisms acquired in the terrestrial environment, and also depend on aerial pollinators to propel sexual reproduction.

A few species, however, developed a peculiar form of sexual reproduction, in which pollination depends exclusively on the aquatic environment. An example of this condition is the Ceratophyllaceae family, which has a single genus Ceratophyllum with a cosmopolitan dispersion.

Species of this genus are entirely submerged, highly specialized to the aquatic habitat, with reduced leaves, without roots and stomata. Its pollination occurs in the water column, forming fruits with hooks that favor their fixation to the muddy substrate and to feathers and legs of aquatic birds.

These characteristics are highly specialized, and taxonomists are often faced with difficulties in recognizing phylogenetic correlations with other plant families. Species of Ceratophyllum do not possess xylem, petals or sepals, which are considered as primitive characteristics, but it is now understood that these adaptations are highly specialized to the aquatic environment. These plants have inhabited the aquatic environment for a long time, even longer than most other aquatic species.

Recent molecular analysis indicated that species of this genus are located in the basis of the angiosperms. Progresses in aquatic plant studies occurred mainly in the last decades of the twentieth century and are continuing, but most of these studies refer to systematics and ecology. Studies related to the phylogenetic correlation and evolution has received little attention until the present.

Understanding the genetic structure of this group, its life history, the meaning of predominance of vegetative reproduction and high frequency of self-fertilization, when compared to terrestrial plants, is still in its nascent stages and is evolving at a slow pace and requires breakthrough advancements in research in order to interpret and explicate all the questions about this issue.

Aquatic plants inhabit swampy or humid environments, but the amount of water accumulated in the environment, and the distribution of the plants depends on topography, geology, and climate.

Wetland and essentially aquatic environments occur in coast regions, deltas, streamside zones, coast or continental lagoons, islands and flooded plains (where half of the wetlands in the planet are found) distributed along the tropical and sub-tropical regions.

Aquatic plants display a wide geographic spread over several continents, and many species are classified as cosmopolites. It is estimated that about 60% of all aquatic plants are in this category, in which the monocotyledons are the most widely distributed group.

The wide distribution of many aquatic species is indicative of the fact that they have efficient and robust mechanisms to disperse their seeds or propagules, enabling the colonization of other lands. Dispersion mechanisms found in this group are transport through water, migrating aquatic birds, wind, and human activities such as navigation.

However, some aquatic plants species may be restricted to a small geographic area. The occurrence of endemism is very low, and generally, it is a result of geographic barriers or specificity for climate or soil type.

There are a few endemic species from swampy areas in Venezuela's mountains and California's seasonal lagoons, in addition to the tropical region of South America, which is particularly rich in endemic species, and the tropics and sub-tropics of Africa and Asia. Among the highly endemic genera are *Sagittaria* and *Echinodorus*, both belonging to *Alismataceae* family.

2.6. MORPHOLOGICAL ADAPTATIONS OF AQUATIC PLANTS

The fauna and flora in water ecosystems is very rich and highly different from one area to another. Some biomes include rainforests, tundra, deserts, taiga, wetlands, rivers, and oceans. Aquatic plants, also referred to as Hydrophytes, have giant leaves and soft bodies to help them allow to float broad or wide.

Even though these plants originated in water, except algae, most of the aquatic plants are evolved from their mesophytic relatives, including crustaceans, aquatic plants such as algae, different species of fish, insects such as dragonflies and mosquitoes, waterfowl, among others.

Figure 2.8: An illustration of hydrilla.

Source: Image by Wikimedia commons.

The morphological adaptations of aquatic plants are:

1. Root system is poorly developed.
2. Roots of floating hydrophytes show very poor development of root hairs, absence of true root caps, with root pockets to protect their tips from injuries (e.g., *Eichhornia*).
3. Rooted hydrophytes like *Hydrilla, Vallisneria*, and *Elodia* derive their nourishment through their body surfaces. More plants partly depend on their roots for the absorption of minerals from the soil. Roots are totally absent in *Ceratophyllum, Salvinia, Azolla, Utricularia,* etc.

4. In *Jussiaea repens* two types of roots develop. Some of them are normal, while others are negatively geotropic, floating roots, spongy in nature and keep the plants afloat.
5. In free-floating hydrophytes, the stem is thick and short, floating on the surface of water (e.g., *Eichhornia*).
6. In *Nymphaea* and *Nelumbium* the stem is a rhizome. These rhizomes live for many years and produce leaves every year.
7. In rooted plants with floating leaves, the leaves are large, flat and entire (e.g., *Nymphaea*, *Victoria regia*). Their upper surface is coated with wax. The wax coating protects the leaves from mechanical and physical injuries and also prevents clogging of stomata by water.
8. In floating plants of *Eichhornia*, *Trapa,* etc., the petioles become characteristically swollen and become spongy, providing buoyancy.
9. Plants such as Limnophylla heterophylla, Sagittaria, Ranunculus, Salvinia, Azolla, etc. show heterophylly, with submerged dissected leaves offering little resistance against the water currents, and absorbing dissolved carbon-di-oxide from water. The aerial leaves show typical mesophytic features. It acts as foliage leaf.
10. Pollination (e.g., *Vallisneria*) and dispersal of fruits and seeds are accomplished by the agency of water.

2.7. PHYSIOLOGY OF AQUATIC PLANTS

In the study of the physiological requirements of aquatic macrophytes, three crucial factors need to be examined; carbon dioxide, light and mineral nutrition. The purpose is to briefly explain a few basic underlying principles of aquatic plant growth, to attain a better understanding of certain practices that will improve the health of aquatic plants.

2.7.1. Carbon Dioxide

Dissolved Inorganic Carbon (DIC) in freshwater occurs as four different species in equilibrium with one another. The four species of Dissolved Inorganic Carbon are; carbon dioxide (CO_2), carbonic acid (H_2 CO_3), bicarbonate (HCO_3^-), and carbonate (CO_3).

The total amount of DIC largely determines the buffering capacity of freshwater, and the ratio of these species with one another largely determines the pH (Wetzel, 1983). Carbon dioxide dissolves readily in water. At air equilibrium, the concentration of CO_2 in air and water is approximately equal at about 0.5 mg/L. Unfortunately, CO_2 diffuses about 10,000 times slower in water than in air (Wetzel, 1983).

This problem is further heightened by the relatively thick unstirred layer (or Prandtl boundary) that surrounds aquatic plant leaves. The unstirred layer in aquatic plants is a layer of still water through which gases and nutrients must diffuse to reach the plant leaf. It is about 0.5 mm thick, which is ten times thicker than in terrestrial plants (Hutchinson, 1975). The result is that approximately 30 mg/L free CO_2 is required to saturate photosynthesis in submerged aquatic plants.

The low diffusivity of CO_2 in water, the relatively thick unstirred layer and the high CO_2 concentration needed to saturate photosynthesis have prompted one scientist to state, "For freshwater submerged aquatic macrophyte plants, the naturally occurring Dissolved Inorganic Carbon levels impose a major limitation on photosynthesis. The Dissolved Inorganic Carbon limitations on aquatic macrophytes and its corollary, the need to conserve carbon, are becoming increasingly apparent as important ecological features of aquatic environments (Bowes, 1985)."

Aquatic plants have adapted to CO_2 limitation in several ways. They have thin, often dissected leaves. This increases the surface to volume ratio and decreases the thickness of the unstirred layer. They have extensive air channels, called aerenchyma, that allow gases to move freely throughout the plants.

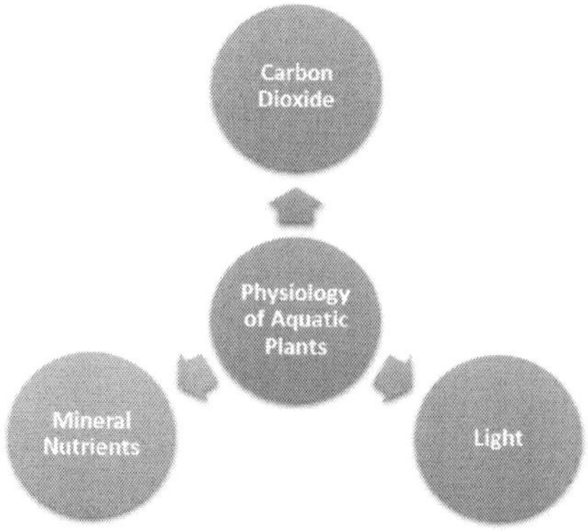

Figure 2.9: Physiology of aquatic plants.

This allows respired CO_2 to be trapped inside the plant and in some species even allows CO_2 from the sediment to diffuse into the leaves. Finally, many species of aquatic plants are able to photosynthesize using bicarbonate as well as CO_2 (Sculthorpe, 1967). This is important, since at pH values between 6.4 and 10.4 the majority of DIC in freshwater exists in the form of bicarbonate (Wetzel, 1983).

2.7.2. Light

Plant chlorophyll absorbs light at wavelengths of 400 to 700 nm. This is termed Photosynthetically Active Radiation (PAR). The intensity of full, natural sunlight is approximately 2,000 umoles/m^2/s, or 100klux, of PAR. Light is attenuated rapidly in freshwater, however, so that submerged aquatic plants receive far less than this amount (Hutchinson, 1975).

Submerged aquatic plants are adapted to the low light levels found in freshwater, and are classified as shade plants on the basis of these adaptations (Sculthorpe, 1967). For instance, aquatic plant chloroplasts, which are the organelles that contain chlorophyll, are often located in the top cell layer of leaves to ensure that as much light as possible is absorbed (Hutchinson, 1975).

Additionally, photosynthesis is saturated at only 15 to 50% full sunlight intensity. Aquatic plants also have a low light compensation point (LCP). The light compensation point is the point at which the rate of photosynthesis equals the rate of respiration and growth stops. This allows them to grow to depths that receive only 1 to 4% full sunlight (20 to 80 umoles/m^2/s PAR).

The attenuation of light in water is wavelength specific. Water absorbs light in the infrared and ultraviolet bands of the spectrum, organic solutes absorb blue, violet and ultraviolet light, phytoplankton absorb blue and orange-red light, and suspended silt absorbs light fairly uniformly at all wavelengths (Wetzel, 1983).

Aquatic plants are therefore exposed to light that is vastly different in quality than incident radiation. Moreover, light quality underwater can change rapidly depending on water depth, turbidity, algal blooms and the level and type of organic solutes present.

These data suggest that aquatic plants are flexible as to their light requirements and that the pursuit of 'full spectrum' light is unnecessary in the freshwater aquarium. There is in fact clear evidence in the scientific literature that freshwater plants can sustain high growth rates under simple cool-white fluorescent light (e.g., Huebert et al., 1990).

Full-spectrum lighting may perhaps be useful, however, for true color rendition, and for attempts by the hobbyist to achieve flowering in 'difficult' aquatic plants. Plants are sensitive to day length. The pigment that senses light in plants is called phytochrome, and it absorbs light in the red/far-red end of the spectrum. Research has shown that some aquatic plants are short-day plants, some are long-day plants, and some are indifferent to daylength.

When exposed to the 'wrong' daylength, plants will continue to photosynthesize in the presence of light, and grow vegetatively, but will not complete their lifecycle and flower. This is true of both terrestrial and aquatic plants.

2.7.3. Mineral Nutrients

Essential mineral nutrients are conveniently separated into two categories. Nutrients used by plants in relatively large amounts are termed macronutrients. They are nitrogen (N), phosphorus (P), sulfur (S), calcium (Ca), magnesium (Mg), and potassium (K). Nutrients used by plants in small amounts are termed micronutrients.

They are iron (Fe), manganese (Mn), copper (Cu), zinc (Zn), molybdenum (Mo), cobalt (Co), and boron (B). Other mineral elements, such as sodium (Na), are also present in plants, but there are at present no definite roles for them, and so they are not classified as essential nutrients.

Aquatic plants, unlike their terrestrial counterparts, can absorb mineral nutrients both from the water through their leaves and from the sediment through their roots. Unfortunately, it is often assumed that rooted aquatic plants can obtain all their mineral nutrient requirements through their leaves.

This is, however, incorrect. As early as 1905, a researcher by the name of Raymond H. Pond stated that, "a soil substratum is requisite for normal growth" and that, "[rooted aquatic plants] make a better growth on a good loam soil, just as many land plants do."

Since then, the dramatic and consistently superior growth of plants rooted in soil compared to plants rooted in sand has been shown repeatedly for many different aquatic plant species from many different types of habitat (Barko and Smart, 1985).

While the reasons for this superior growth are not completely understood, certain facts are clear. First, submerged soils are generally lacking in oxygen (Wetzel, 1983). This is of benefit to rooted aquatic plants since under anoxic conditions Fe, P and N are more readily available than under aerobic conditions. Second, nutrient concentrations are higher in a fertile soil than in the overlying water.

Third, there is no competition with phytoplankton for available nutrients. This latter point is critical, since with water-based nutrition, an excess of fertilizer will cause phytoplankton blooms.

Rooted aquatic plants are well adapted to growing in an anaerobic substrate. They are able to 'pump' enough oxygen to the roots so that in many cases the oxygen actually diffuses into the surrounding sediment (Sculthorpe, 1967).

They can also respire anaerobically if necessary and produce lactic acid or ethanol instead of CO_2 as a byproduct. The root meristems (growing tips) of some species are even inhibited in the presence of oxygen (Sculthorpe, 1967).

Aquatic plants also have requirements for certain nutrients in the overlying water. Most rooted aquatic plants need Ca, Mg, K, and a carbon source in the water if they are to thrive (Barko and Smart, 1981). Since some aquatic species such as *Isoetes* sps., and *Lobelia dortmanna* actually obtain even their carbon dioxide from the sediment.

These plants are adapted to growing in acidic soft water lakes that have extremely low levels of Dissolved Inorganic Carbon in the water, and so absorb CO_2 from the sediment through their roots (Keeley, 1982; Boston et al., 1987).

It can be established thereby that aquatic plants are superbly adapted to their underwater habitat. Many are in fact considered serious weeds in aquatic ecosystems throughout the world. The capacity of aquatic plants to grow and fill space can be truly remarkable, and a doubling of plant biomass in 2 to 4 days is not unusual under optimal conditions.

2.8. CELLULAR BIOLOGY OF AQUATIC PLANTS

All cells consist of a cell membrane that separates the inside of the cell from the outside environment. The cell membrane controls the movement of materials into and out of the cell. The cell membrane is composed of a double layer of phospholipids. Phospholipids are organic molecules that have a non-polar hydrophobic (water-avoiding) end and a polar hydrophilic (water-attracting) end.

These two layers of phospholipids form a bilayer "sandwich." The hydrophobic ends face each other, forming the sandwich "filling." The hydrophilic ends form the sandwich "bread" on each side of the membrane. Hydrophilic ends thus have contact with the inside and outside of the cell.

The phospholipid cell membrane also has proteins embedded in it. The proteins have various functions. Some proteins act as enzymes to catalyze biological reactions, and other proteins act as channels for transporting molecules in and out of the cell.

Cell walls surround the cell membrane and provide protection and structural stability for the organism. Although the cell wall can be rigid, it is usually flexible and allows the diffusion of water and nutrients. Cell walls are present in the cells of eukaryotic plants, algae, and fungi.

2.8.1. Cellular Water Balance

Cell membranes permit a cell to regulate its internal environment. Water and other molecules pass across the cell membrane through diffusion. Diffusion is the tendency for molecules of any substance to spread out into the available space. Diffusion is driven by thermal energy, which results in movement through random vibration of molecules.

Molecules generally travel from areas of high concentration to areas of low concentration and across permeable or semi-permeable barriers. The cell membrane is selectively permeable to molecules of a certain size and charge, meaning some substances pass through the lipid bilayer freely, whereas others must be transported.

Non-polar molecules like carbon dioxide (CO_2) and oxygen gas (O_2) pass through the cell membrane easily through diffusion. Water (H_2O) can also easily cross the membrane. Even though it is a polar molecule, the small size of a water molecule allows it to diffuse across the membrane. Contrastingly, large polar molecules like glucose have big hydrophobic sections.

Their large size prevents them from passively diffusing across the membrane. In order to pass through, large, polar molecules must be actively transported. Active transport uses energy stored by the cell and is accomplished by proteins embedded in the cell membrane.

The concentration of dissolved substances is often very different inside a cell compared with the outside environment. For example, the inside of a cell tends to be higher in proteins than the outside. The inside of cells also contains higher concentrations of nutrients and dissolved gases necessary for respiration, growth, and reproduction.

The external environment of a cell also contains solutes. Solutes are substances dissolved in fluids, like table salt, which dissolves in water as sodium and chloride. A solution is a fluid that contains dissolved solutes.

Consider a cell placed in seawater, which contains several dissolved salts and other trace elements. The difference in relative concentration of dissolved substances inside and outside the cell has very important consequences for cell size.

Water will diffuse across a membrane from an area of low solute concentration to an area of high solute concentration through a process called Osmosis. Osmosis is the special term given to the diffusion of water across a permeable surface. Water will move until the solute concentration on both sides of the membrane is equal.

In contrast to animal cells, algae and plant cells have cell walls. A hypotonic solution can cause the cell to become very firm or "turgid" because the rigid cell wall pushes back as water flows into the cells. This keeps the entire plant erect, allowing for tall plant stems and upward growth. The rigid structure of the cell wall actually prevents over-expansion.

Plant cells placed into isotonic solutions bend down. Even though the solute concentrations are equal, plants wilt and become flaccid, without the turgor pressure of the cell wall to stay firm.

In the case of a hypertonic solution, water leaves the cell. The cell membrane will pull away from the cell wall. Plasmolysis is the term used to describe the process of cells losing water in a hypertonic solution.

Osmosis, as described above in blood and plant cells, is a passive movement of water. It is important to note that some organisms also actively transport water into or out of the cell. This process is called osmoregulation, and it requires energy input by the organism.

2.8.2. Cellular Components

Organelles are specialized subunits within the cell. Organelles are usually enclosed in another lipid membrane. The nucleus is an organelle found in eukaryotes that contains the genetic material of the cell and controls cell processes. The interior cell structures are surrounded by cytoplasm. Cytoplasm is a gel like substance made mostly of water.

Organelles compartmentalize cell tasks. Eukaryotes have several important organelles. For example,

1. Mitochondria are known as the "cellular power plants" because they supply much of the cell's chemical energy.
2. Ribosomes and the rough endoplasmic reticulum are organelles responsible for protein synthesis.
3. Smooth endoplasmic reticulum is responsible for metabolism and detoxification.
4. Golgi apparatus is composed of stacks of membranes and is responsible for packaging and sorting proteins made within the cell.

Eukaryotic cells contain a membrane-bound nucleus and other organelles. Their cells are range in size from 10 to 100 micrometers (µm). Eukaryotic organisms can be unicellular or multicellular. The nucleus of eukaryotic cells contains at least two chromosomes. A chromosome is a single piece of DNA that contains many genes.

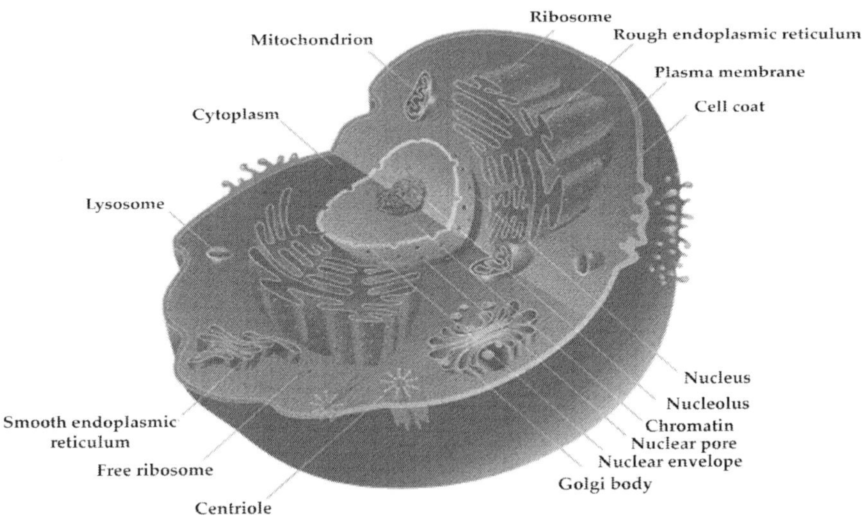

Figure 2.10: Diagram of eukaryotic cell.

Source: Image by Wikimedia commons.

Prokaryotes, on the other hand, have much smaller cells (1–10 µm) and are only unicellular. Prokaryotes do not have a nucleus or any other membrane-bound organelles. In prokaryotes, their DNA is contained within a single, circular chromosome inside the cytoplasm.

2.8.3. Cells of Photo-Synthesizers

All photosynthetic autotrophs have structures inside their cells that harvest and convert sunlight into usable energy. However, there are differences between prokaryotic and eukaryotic cells in terms of where photosynthesis occurs.

Prokaryotic autotrophs perform photosynthesis and respiration within thylakoid membranes. Thylakoid membranes are generally located near the outer cell membrane and contain pigments like chlorophyll that are responsible for photosynthesis.

Eukaryotic autotrophs are more complex. Most eukaryotic autotrophs have organelles called chloroplasts that contain the machinery for photosynthesis. The inside of the chloroplast is filled with stroma, a watery substance similar to cytoplasm. Thylakoids containing the photosynthetic pigments are located within the stroma. Grana are arranged stacks of disc-shaped thylakoids.

Interestingly, cyanobacteria are considered the ancestors of modern chloroplasts. Scientists think cyanobacteria were incorporated into eukaryotic cells through a symbiotic relationship. A free-living cyanobacteria was likely taken inside the eukaryotic cell as a helpful organism and eventually became part of the host. This same process also likely occurred for eukaryotic mitochondria.

There is evidence for chloroplasts and mitochondria previously being separate organisms:

1. Both contain their own DNA. The DNA in chloroplasts and mitochondria is separate from the DNA in the cell's nucleus.
2. The chloroplast and mitochondria organelles also have two membranes. The second membrane is thought to have been produced when the host's cell membrane surrounded a cyanobacterium as it entered the host cell.

2.8.4. Plant Structure

Plants are photosynthetic organisms. Almost all plants have vein-like vascular tissue and structural systems that transport water and nutrients throughout the plant. There are some plants without vascular tissues—moss-like species that roughly resemble some seaweeds or algae. This section describes vein-like vascular plants.

The body plan of plants is generally broken down into three systems: the leaves, the root, and the shoot. The leaves of the plant are the primary location of photosynthesis. Roots provide an anchoring system for the plant, in addition to absorption.

Figure 2.11: An illustration of lato seaweeds.

Source: Image by Flickr.

For most plants, water and nutrients are absorbed through the roots. The shoot system provides a method of exchanging water and nutrients between the roots and the leaves through a network of vascular tissue. The shoot system also allows plants to grow vertically. Sunlight can be a limiting resource if the plant is shaded, so having the ability to grow taller than plant neighbors is an important adaptation for plants.

2.9. CONCLUSION

Aquatic plants, or Hydrophytes, have not just adapted to living in an aquatic environment, but they are also the primary producers and are the foundation of the food web for plenty of organisms. They have a relevant impact on the soil chemistry and light levels since they reduce the flow of water and also catch pollutants and sediments.

As it was observed in the chapter that aquatic plants require special adaptations for either living submerged or afloat water. The most relevant adaptation is the presence of lightweight internal packing cells, aerenchyma, but floating leaves and finely dissected leaves are also common. Aquatic plants can only grow in water or in soil that is frequently saturated with water.

Due to their underwater environment as in the case of submerged plants, photosynthesis is also a complex process. Yet, in spite of these shortcomings, aquatic plants have developed and adapted various mechanisms to maximize absorption.

Aquatic plants perform many important functions in the aquatic ecosystems, i.e., dissolving nutrients from water, constructing wetlands to clean polluted water bodies, sedimentation of suspended solids in the water by reducing the velocity, and lower erosion levels by stabilizing soil surfaces.

Conservation of aquatic plants is of utmost importance as they are also used by humans as food, for example, include wild rice (*Zizania*), water caltrop (*Trapa natans*), Chinese water chestnut (*Eleocharis dulcis*), Indian lotus (*Nelumbo nucifera*), water spinach (*Ipomoea aquatica*), and watercress (*Rorippa nasturtium-aquaticum*).

REFERENCES

1. Baker, A. L. (n.d.). Phycokey – Aquatic Macrophytes. [online] Cfb.unh.edu. Available at: http://cfb.unh.edu/phycokey/Choices/Anomalous_Items/Aquatic_macrophytes/Aq_macrophytes_key.html (accessed on 31 July 2021).
2. Barrett, S., Eckert, C., & Husband, B., (1992). Evolutionary processes in aquatic plant populations. *Aquatic Botany*, [online] *44*(2–3), 105–145. Available at: https://www.sciencedirect.com/science/article/abs/pii/0304377093900688 (accessed on 31 July 2021).
3. Billet, K., Genitoni, J., Bozec, M., Renault, D., & Barloy, D., (2018). Aquatic and terrestrial morphotypes of the aquatic invasive plant, Ludwigia grandiflora, show distinct morphological and metabolomic responses. *Ecology and Evolution*, [online] *8*(5), 2568–2579. Available at: https://www.ncbi.nlm.nih.gov/pmc/articles/PMC5838032/ (accessed on 31 July 2021).
4. Du, Z., & Wang, Q., (2015). Phylogenetic tree of vascular plants reveals the origins of aquatic angiosperms. *Journal of Systematics and Evolution*, [online] *54*(4), 342–348. Available at: https://onlinelibrary.wiley.com/doi/full/10.1111/jse.12182#:~:text=Aquatic%20lineages%20evolved%20early%20in,habitats%20during%20the%20Early%20Cretaceous. (accessed on 31 July 2021).
5. Edna Scremin-Dias, (n.d.). Tropical Biology and Conservation Management. Vol. I: Tropical Aquatic Plants: Morphoanatomical Adaptations. [ebook] *Encyclopedia of Life Support Systems (EOLSS).* Available at: https://www.eolss.net/Sample-Chapters/C20/E6–142-NH-02.pdf (accessed on 31 July 2021).
6. Encyclopedia.com, (2019). Plant Anatomy | *Encyclopedia.com.* [online] Encyclopedia.com. Available at: https://www.encyclopedia.com/social-sciences/applied-and-social-sciences-magazines/plant-anatomy (accessed on 31 July 2021).
7. Kanchanaree Pongchawee, (n.d.). Advanced Freshwater Aquaculture: Aquatic Plants in Aquaculture. [ebook] Available at: https://sswm.info/sites/default/files/reference_attachments/PONGCHAWEE%20ny%20Aquatic%20plants%20in%20Aquaculture.pdf (accessed on 31 July 2021).
8. Penny Porter, (2017). Fixed Aquatic Plants. [online] Sciencing. Available at: https://sciencing.com/fixed-aquatic-plants-12404457.html (accessed on 31 July 2021).

9. ProFlowers, (2016). Aquatic Plants and Flowers. [online] proflowers.com. Available at: https://www.proflowers.com/blog/aquatic-plants-and-flowers#:~:text=Water%20hawthorns%20(apogeton%20distachys)%20are,with%20fish%20living%20in%20it. (accessed on 31 July 2021).
10. University of Hawaii, (2021). Structure and Function | www.manoa.hawaii.edu. *Exploring Our Fluid Earth.* [online] Manoa.hawaii.edu. Available at: https://manoa.hawaii.edu/exploringourfluidearth/biological/aquatic-plants-and-algae/structure-and-function (accessed on 31 July 2021).

Chapter

3

Advances in Biofuel Production: Algae and Aquatic Plants

CONTENTS

This chapter revolves around the advances in biofuel production, as all the benefits of the microalgal biofuels have been discussed along with the challenges related to algal fuel commercialization. And after this, the discussion is all about the algae-based bioenergy products and for that what technologies have been adapted or being considered and also the threats which are there for the developing countries in algae-based biofuels. After the selection of technology, what role and importance the technology is having in the biodiesel production is also being discussed and how the aquatic plant Azolla is the universal feedback for the biofuel production.

3.1. INTRODUCTION

To perform the function, the global economy requires the fossil hydrocarbons, from producing plastics and fertilizers, as for lighting, heating, and transportation it up the energy is being required. There will be increased in the use of fossil fuel along with the increase in population and expanding economy. As per the data, it has been suggested that the use of fossil fuel use will increase, and competition for these limited resources will increase as countries improve their gross domestic product per capita. Moreover, there will also be an increase in the atmospheric CO_2 concentration, and the potential for the significant greenhouse gas-mediated climate change, which will affect all parts of the world. In the end, it has been seen that it is being partially derived from the ancient algae deposit and to recover it is a limited resource that will eventually run out or become too expensive. The development of the renewable energy sources is being driven by these factors that can supplant fossil fuels, and for all the nations, it allows greater access to fuel resources, while greatly reducing carbon emissions into the atmosphere.

Figure 3.1: Algae can be used in the production of biofuel.

Source: Image by Wikimedia Commons.

As the renewable energy sources a number of technologies have been examined and to provide a total solution there is no single strategy available, and it seems possible that a combination of strategies can be employed that will substantially decrease our dependence on fossil fuels. With the existing energy options, it can be cost-competitive and the challenge possible that a combination of strategies can be employed that will substantially decrease our dependence on fossil fuels.

For the generation of electrical power, fossil fuels are being used as well as liquid fuels. To generate the electrical power, there are a variety of renewable or low atmospheric pollution technologies, and it also includes solar, wind, hydroelectric, geothermal, and nuclear.

Though, it has been seen those renewable technologies to supplement or replace liquid fossil fuels are still in their early developmental stages. By 2030 the international energy agency expects that biofuels will contribute 6% of total fuel use, but if undeveloped petroleum fields are not accessed or if substantial new fields are not identified, it could expand significantly.

Under the moniker biofuels, the most promising sustainable alternatives are almost exclusively categorized. A diverse range of the technologies that generate the fuel is being described by this term with at least one component based on a biological system.

For the biofuels, the major technologies presently employed begins with the terrestrial plants and culminate with ethanol, whether this is corn starch to sugar to ethanol, or sugarcane sugars to ethanol. In particular, it has been said that the regional success of some of these strategies is well noted that the sugarcane-to-ethanol production in Brazil. From the terrestrial plants there is a lesser degree oils for example, soy and palm which are being used to produce biodiesel.

At the small scale, these strategies are functional, though as their use has increased and it is being evident that they are not sustainable and in order to supplement a significant fraction of petroleum using this strategy there is a need for owing to the enormous amount of agricultural land that would be required. There are a number of hybrid strategies which have been discussed or are currently being deployed.

For the fermentation into the fuel, there are examples of such strategies include conversion of cellulose to sugars and to produce liquid fuels the gasification of residual biomass into syngas that can then be used. Though, to produce fuels, each of these strategies being used and to accommodate the global demand for liquid fuels are insufficient.

3.2. BENEFITS OF MICROALGAL BIOFUELS

To offer a variety of solutions, there are microalgae which are the diverse group of single-celled organisms which are having the potential for the liquid transportation fuel requirements through a number of avenues. In the wide range of aquatic environments, algal species are been grown through the saturated saline from freshwater.

For more than 40% of the global carbon fixation algae efficiently use CO_2 and are also responsible, with the majority of this productivity coming from marine microalgae. With some species doubling algae can produce biomass rapidly, in as few as 6 h, and many exhibiting two doublings per day.

To produce energy rich oils, all the algae have the capacity, and a number of microalgal species have been found to naturally accumulate high oil levels in total dry biomass. For example, there are some *Butyraceous* spp. have been identified that have up to 50% of their dry mass stored as long-chain hydrocarbons.

For identifying the production strains, algal diversity gives researchers many options with potentially millions of species, and for genetic information, it also provides sources that can be used to improve these production strains.

From the groups, there are the microalgal species which is being investigated as potential biofuel crops originate whose ancestral relationships are significantly broader than the most diverse land plants, providing a wealth of genetic diversity.

At the time of discussion about the microalgae, which are the diatoms, green algae, golden brown, rhyniophytes, eustigmatophytes, and cyanobacteria, and the groups most often being considered along with the members from all of these groups have been examined as potential fuel production strains. Though, cyanobacteria are not algae but a class of photosynthetic bacteria which is to be noted.

Over the terrestrial plant's microalgae have the additional advantages. By the division, it has been seen that they are single-celled organisms that duplicate, and high-throughput technologies can be used to rapidly evolve strains.

In the crop plants, this can reduce processes that take years, down to a few months in algae. On the environment, algae have a reduced impact compared with the terrestrial sources of biomass which is being used for biofuels. For the traditional agricultural, they can be grown on the land which will not be used and are very efficient at removing nutrients from the water.

Therefore, from the terrestrial plants, not only would production of algae biofuels minimize land use compared with biofuels produced but, in the process of culturing these microalgae, waste streams can be remediated.

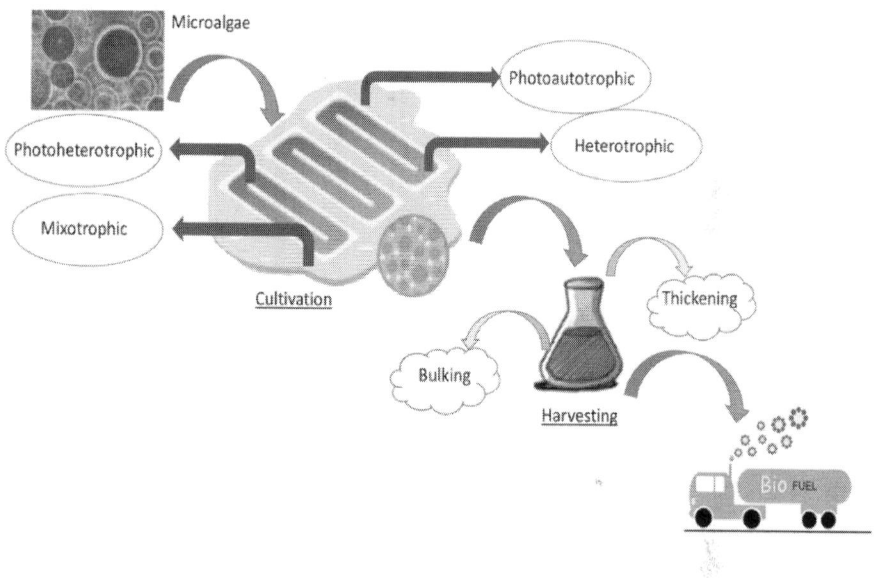

Figure 3.2: An illustration of production of biofuel using microalgae.

Source: Image by Wikimedia Commons.

In order to remove nitrates and phosphates before discharge potential waste streams include municipal wastewater, and to capture sulfates and CO_2, there is a need for flue gas of coal or other combustible-based power plants.

To be bioengineered algae production strains also have the potential, and along with this, it allows improvement of specific traits and production of valuable co-products, which may allow algal biofuels to compete economically with petroleum. With a high potential, these characteristics make algae a platform to produce cost-competitive biofuels.

3.3. CHALLENGES FOR ALGAL FUEL COMMERCIALIZATION

To invest significant capital to turn algae into biofuels, the high growth rates, reasonable growth densities, and high oil contents have all been cited as reasons. Therefore, to offset petroleum, there is a need for algae to mature as an economically viable platform and consequently, mitigate CO_2 release, there are a number of hurdles to overcome, ranging from how and where to grow these algae, to improving oil extraction and fuel processing.

It has been seen that algal biofuels production chain is being outlined, and it shows that the major challenges which include strain isolation, nutrient sourcing and utilization, production management, harvesting, coproduct development, fuel extraction, refining, and residual biomass utilization.

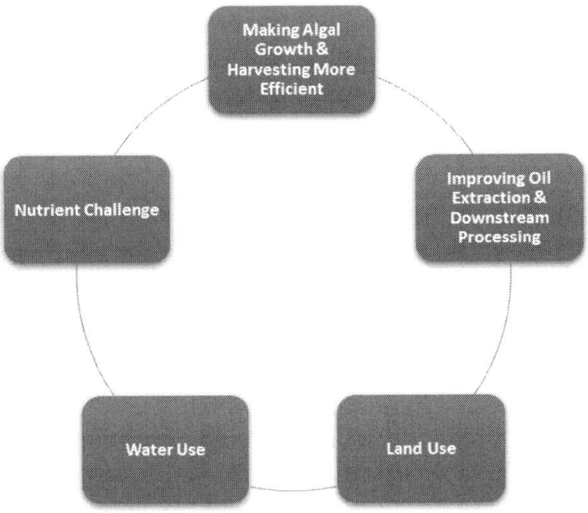

Figure 3.3: Challenges for algal fuel commercialization.

3.3.1. Making Algal Growth and Harvesting More Efficient

On the algae biofuel production, improved engineering will make a significant impact. For the nutrient circulation and light exposure, these improvements include efficient strategies, and which have been reviewed elsewhere.

For the engineers, there are significant challenges to design the photobioreactors (PBRs) that are cheap enough for large-scale deployment, or for engineers and biologists to combine forces to develop species that grow efficiently in low-cost open systems. Over open systems, PBRs have the advantages in that they can more easily maintain axenic cultures, and in order to make an increase in the productivity, it can maintain more controlled growth environments, though by efficiencies in gas exchange contained systems are being challenged and also a requirement for supplemental cooling.

Figure 3.4: An illustration of photobioreactor.

Source: Image by Wikimedia Commons.

As it is unclear whether PBRs will ever become cost-competitive with open pond systems, despite the advantages of decreased contamination and increased productivity. For the growth, harvesting and extracting oil, there is the substantial improvements over the current technologies regardless of the growth strategy being employed.

3.3.2. Improving Oil Extraction and Downstream Processing

From the engineering side, oil extraction is another challenge that is most easily addressed. For extracting the oil from algae, there are three major strategies such as oil press/expeller, hexane extraction, and supercritical CO_2 fluid extraction.

These technologies are relatively expensive and also have been successfully demonstrated, either in terms of equipment needed or energy required to extract the oil. Fortunately, to the engineering improvements all are amenable. To the crude fossil fuel oil, crude algae oil is chemically similar once it is being extracted, and the engineering challenges associated with algae oil conversion to usable liquid fuels are similar to those already well managed by petroleum companies, although improved catalysts will be required to improve gasoline production from bio-oil.

Figure 3.5: Improving oil extraction and downstream processing.

Source: Image by Flickr.

And only because of these similarities it seems reasonable to assume that collaborations between algae production companies and major oil companies are likely, since these companies have extensive experience maximizing downstream processing efficiencies.

3.3.3. Land Use

To replace a meaningful amount of fossil fuel is significant the scale of implementation that is required, regardless of the growth strategy employed and efficiency of oil extraction. It has been said that per day in 2008, the USA alone required 19,497,950 barrels of oil.

To impact this number for the algae or any other biofuel feedstock there is the significant acreage must be dedicated to production facilities, with estimates suggesting that 30 million acres will be required to meet US oil demand. For the process of how large-scale aquaculture can be achieved, different models have been presented.

So, in this article focus is on the terrestrial aquaculture, although both terrestrial strategies and marine strategies may be required since marine strategies are completely unknown at present and may require engineering significantly different from what is practiced today. For the food agriculture, the terrestrial models use land that is not presently used and has minimal known environmental or other significant economic utility.

3.3.4. Water Use

In the algal growth water is potentially a major limiting factor. Into the nonarable land, the expansion of algal growth will require water, fortunately in many of these regions which have the substantial alkaline or saline water reservoirs beneath them by providing a significant source of nonportable water that is suitable for growth of many algal species.

It has been said that to replenish the water lost in evaporation, algae grown in open ponds have water requirements per unit area similar to that of cotton or wheat but less than that of corn. To consider water use to avoid a future it is imperative when considering broad deployment of algae for water versus fuel. Though, for algae biofuels production, substantial alkaline reserves are available, and water will remain a central issue and which is need to be considered carefully as the industry expands.

3.3.5. Nutrient Challenge

For the efficient growth, algae require nutrients, light, water, and a carbon source, most often CO_2. It has been seen that the major nutrients required by most algae include phosphorous, nitrogen, iron, and sulfur. As it has been seen that nutrient requirement necessary for algal growth is ignored, since algae are very efficient at sequestering these nutrients when present in their environment. In terms of eutrophication of lakes and coastal regions, change in nutrient load and algal growth have been studied extensively but not as heavily in terms of productivity in large-scale aquaculture. For some of the challenges if terrestrial agriculture is a model for algal aquaculture, and there is significant challenge providing sufficient nutrients for large-scale algal growth.

In the current terrestrial agriculture industry micro and macronutrient supplements, or fertilizer, account for significant costs and biofuels are not expected to be an exception. Globally it has been seen that the use of fertilizers has been increasing. Unfortunately, from the fossil fuels or mined many fertilizer components are generated, and they are not renewable.

The major components of the agricultural fertilizers are similar to plants which require sources of phosphorus, nitrogen, and potassium, and large e-scale aquaculture will impact these already limited supplies. Moreover, it has been said that chelated iron and sulfur is being required by the optimal growth of many algal species. It has been seen that less than 1% of total algal biomass which the phosphorus makes up slightly and which is required at approximately 0.03–0.06% in the medium to sustain algal growth. In the USA, the fertilizers which are being used for the agriculture currently contain a less than optimal concentration of phosphate owing to limited supplies.

From the USA, less than 40 million tons of phosphate is mined annually and the maximum phosphate production from this mining peaked in the late 1980s. In the USA, there is a situation if algal biofuels are to completely replace petroleum, then an additional 53 million tons of phosphate must be acquired annually. So, in the USA, this is considered as the significant challenge that the total amount of phosphate is estimated to be approximately 2.8 billion tons. In supply, the nitrogen is not limited unlike the phosphorus but when it comes to plant and algae growth it is often a limiting macronutrient. Into the ammonia, nitrates, and similar molecules algae require nitrogen to be fixed in order to be used as a nutrient source.

It has been seen that there are some of the bacteria such as rhizobia which are having the ability to fix their own nitrogen and some form symbiotic relationships with terrestrial plants, providing the plants with this crucial nutrient to sustain protein and nucleic acid synthesis.

To fix the nitrogen, some cyanobacteria also have the ability, while there is a requirement for an exogenous source of fixed nitrogen to date and almost all algal species identified and most prefer ammonia, as it is less energetically demanding than nitrate or nitrite.

For the algae biofuel production there is need for providing a cheap source of fixed nitrogen, as it will be important and to supply this nitrogen there is the possibility of using nitrogen-fixing which may help in minimizing these costs.

3.4. ALGAE-BASED BIOENERGY PRODUCTS

To produce biofuel with algae, there are a variety of way. So, in this section, to produce various biofuels, only the requirements of the algal biomass are needed and are briefly discussed in order to facilitate the selection of co-production options further in the report.

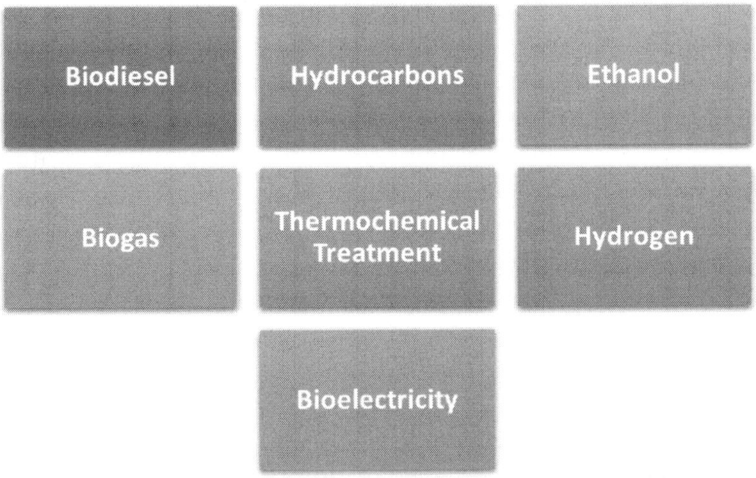

Figure 3.6: Algae-based products.

3.4.1. Biodiesel

From the algal oils' biodiesel production has received the most attention since algae can contain potentially over 80% total lipids. With the very low yields under the normal growth conditions, the lipid concentration is lower (<40%), and high oil content is always associated. Under such condition's biomass production is not optimal though, reducing the non-lipid part of the biomass that can be further used as a source for co-products.

3.4.2. Hydrocarbons

It has been said that there is one genus of algae that does not produce the above-mentioned lipids, but apart from this, it produces longer chain hydrocarbons, which are not suitable for biodiesel production. And there is also a suggestion that under a process that is similar to the production of conventional fuels from fossil oil they can be converted. In the saline water, it has been seen that Botryococcus, which is a cross water species can be grown, and it can also produce certain carotenoids. Slow growth speed is the relatively its drawback.

3.4.3. Ethanol

From the starch, ethanol is commonly produced, which also contain the feedstocks, as some algae have been reported to contain over 50% of starch.

in a process similar to cellulosic ethanol production, algal cell walls consist of polysaccharides which can be used as a feedstock, along with the added advantage that algae rarely contain lignin and their polysaccharides are generally more easily broken down than woody biomass. From the non-carbohydrate co products can potentially be derived as a part of the algal biomass.

3.4.4. Biogas

Into the biogas anaerobic digestion converts organic material that contains about 60–70% biomethane, while the rest is mainly CO_2, which can be fed back to the algae. As reducing the need for drying, a main advantage is that this process can use wet biomass. From the liquid and solid phase, another advantage that is being seen is that the nutrients contained in the digested biomass can be recovered.

Figure 3.7: An illustration of biogas plant.

Source: Image by Wikimedia Commons.

3.4.5. Thermochemical Treatment

There is an advantage that no live organisms are involved with the biological treatment of algal organic material which has a non-biological counterpart, and therefore more varied and extreme process conditions can be used.

Under the high temperature and pressure conditions, the biomass undergoes a chemical conversion. How extreme these conditions are is depending on the water content, in a raw gaseous, liquid, or solid phase, the biomass carbons end up which can be upgraded for usage as a biofuel. As compared to the biogas production, the energy input of this type of treatment.

3.4.6. Hydrogen

Into producing the hydrogen gas, some algae can be manipulated. Since the energy is lost by the cells to form hydrogen is being lost currently the yield of this process is low and not much biomass is produced and therefore there is little potential for co-production.

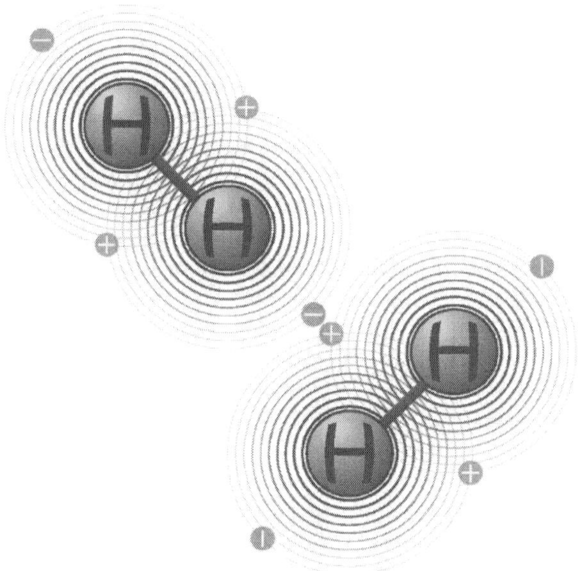

Figure 3.8: An illustration of hydrogen molecule.

Source: Image by Flickr.

3.4.7. Bioelectricity

In a power plant, algal biomass can also be co-combusted. And it has been said that for this biomass needs to be dried, which implies a significant amount of energy. So, to extract a certain co-product as a first step before being used as a biofuel, then there should be a situation that if the biomass is required to be dried in order and then only the process is only interesting.

3.5. TECHNOLOGICAL OPPORTUNITIES AND THREATS FOR DEVELOPING COUNTRIES IN ALGAE-BASED BIOFUELS

So, it has been seen that most of the algae concepts are immature, and most of the technological barriers are fundamental and of global relevance. Though, in developing countries, there are some of the socio-economic and geographical aspects present which is leading to both opportunities and barriers. Firstly, it has been seen that in the developing countries, food security is of importance, and to tackle this algae concepts (co)-producing food or feed provide the opportunities.

In addition to it there some of the developing countries which are being found at the latitudes with high annual solar radiation and there is a key to a high productivity and with the less sunshine it may also attract investments from richer foreign regions. In the developing countries, another attraction for the investment is the lower wages and the lower costs for land and some required inputs and construction materials.

Therefore, locally some parts of the material may not be available and because of these expensive imports are being required. And specially among the poor, this product in the local market is based on the lowest possible price, and on the other hand, in the developed countries, a healthy or "green" product may be sold at a higher price.

Figure 3.9: Technological opportunities and threats for developing countries in algae-based biofuels.

Source: Image by Flickr.

In the developing countries introducing and publicizing a new product is more difficult. For the poor independence of foreign oil/energy and energy access will greatly help both the economy and raising living standards. In addition to it, in developing countries, some of the negative impacts associated with plant-based biofuels are avoided by using algae.

3.5.1. Scale of Operations

On the scale of operations, the economics of algae-based biofuel production are often heavily dependent. Large-scale facilities can achieve a lower production price/kilogram biomass through economies of scale. It has been said that for an algal biodiesel plant is 1000 ha, which pumps about 1 billion liters of saltwater a day.

It has been said that for the developing countries, there are two scenarios which are foreseeable, one where these kinds of scales are achieved, and one where the concepts are reduced to the community level of the rural poor. For the input and output, the large-scale scenario requires large investment and market, as for construction, operation and management there is required for sufficient skilled personnel.

This mean in an urban setting, such a concept will be more viable with substantial industrial development. In the developing countries, the current commercial example of algae cultivation fit in this category. On a very large-scale algae farming may result in alienation and lack of integration between the environment and people.

For their possible environmental and social impacts, these projects should be analyzed thoroughly. Into the migration, the main risk is forcing human population and undervalue cultural and religious attachment to the land that contributes to well-being, destroys or disrupts entire ecosystems and their inhabitants and animals.

A strong effort needs to be made to integrate them into the existing ecosystem and social system only if such large projects are envisioned. If the technology doesn't perform as expected, then the risk will be higher, as the larger the system. In the case, if projects are well being designed the positive impacts can be expected.

The small-scale community operated system is much more appropriate for the one billion rural poor. For this, there is one more consequence that the initial investment costs will generally be a more significant barrier than in large industrial projects where the long-term profitability is being pursued. And the open pond system is must cheaper to construct that's the reason they more likely choice. For the cultivation open system limit the species which are available. Both nutrients and CO_2 are essential to obtain sufficient productivity. From the waste streams, low-cost nutrients will generally be available, and for the energy generation CO_2 supply may require nearby continuous (bio or fossil) fuel combustion, for instance.

In addition to it, in technology harvesting requires significant investment which can be consider as another crucial barrier. There are two options by which the need for expensive harvesting technologies can be avoided. First is cultivating filamentous (thread-forming) species of algae like Spirulina and the second one is feeding live algae to fish (or algae-eating organisms that serve as fish feed).

A protein rich food source is being primarily provided by the both concepts. In such a system the most likely option for co-producing bioenergy is digestion, which allows co-digestion of other organic waste streams, recycling of nutrients into the algae cultivation system and provides biogas, which can be used for cooking, heating and lighting or on a larger scale for electricity generation which feeds its CO_2 emissions into algaculture.

In the sun, it is possible to dry harvested algae naturally for the small systems, while large operations will focus on using all available land to capture sunlight for algae cultivation. It has been seen that the oil can be relatively easily extracted from the dry biomass using an oil press similar to the ones used in manual soybean oil extraction only if the oil rich algae can be cultivated.

The good animal feed will be made by the left-over biomass. From the reduced logistics cost the economies of small-scale systems are also being benefited. For example, it was determined that for South African biodiesel plants, the increased cost of production due to higher capital cost per unit should be more than offset by savings in transportation cost.

3.5.2. Potentials for Algae Production: Limitations from Water Requirements

On the wellbeing, shortage of water can lead to heavy impact and water is a limited resource, which is being possibly forced migration and episodes of famine. In addition to it, to exacerbate existing issues, climate change is likely.

Areas which are being located in the sunny regions small-scale systems will likely be open, shallow and a large amount of water will be lost through evaporation. In the arid regions, this severely restricts the possibilities, unless an alternative water source is available, but also regions with high annual rainfall may experience dry and wet seasons.

From the urban areas, alternative water sources may be found like wastewater streams or, in some cases, groundwater is available, but the cost of pumping the water to the cultivation system may be too high.

Worldwide it has been seen that due to the scarcity of freshwater reserves and the unsustainable use of freshwater aquifers, large operations should only consider the use of brackish water or seawater. It can be well explained with the example from the US department of energy calculates 60–454 trillion liters of saline water use per year to displace diesel use in the United States with algae diesel, depending on achievable productivity.

In the United States, the current saline aquifer extraction is approximately 83 trillion liters, while freshwater use for US corn cultivation is upwards of 15,141 trillion liters per year. For the algal growth is substantial water consumption is still favorable compared with agricultural crop production.

3.6. ALGAE-BASED NON-ENERGY OPTIONS

For the number of products, the application of algae is practically infinite, by the different cultivation systems, large diversity and changes in chemical composition is being influenced. Due to the entrapment of this natural resource, the field of algae-based products is very narrow.

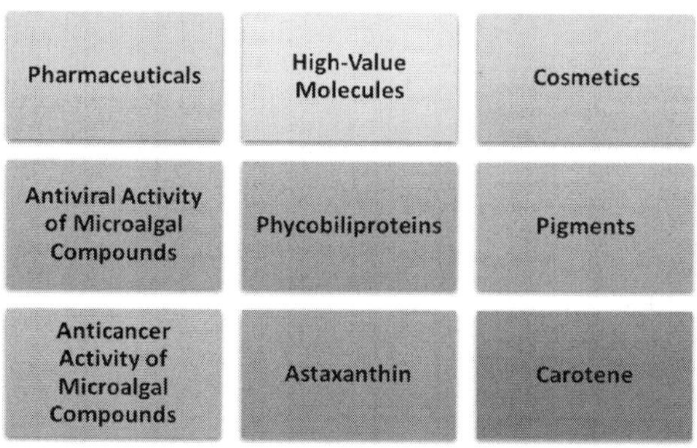

Figure 3.10: Algae-based non-energy options.

For this economic limit, there are many factors like approval of new products by regulating authorities, non-awareness about products, lack of knowledge for algae cultivation, and lack of investments for establishing large scale facilities. From the algae bulk of commercial products are available, most products as food, alginates are derivatives of seaweed and moderately derived from natural populations than cultivated.

3.6.1. Pharmaceuticals

Including the primary and secondary metabolites algae are the rich sources of peculiar biologically active compounds which could be used as a potential candidate of interest for the pharmaceutical sector. Due to the inhabitation of these organisms in natural aquatic communities, the presence of various bioactive compounds in algae is anticipated, and in the same niche where an inhibitory competition occurs between producers and consumers.

For the commercial applications, microalgae are the source of immense bioactive compounds that can be exploited. A wide range of pharmacy products, proteins, vaccines, nutrients is being provided by the algae and from animal and plant sources that otherwise are not available or are very costly to produce. From the microalgae, various pharmaceutical products have high value, but its commercialization is still in its infancy and but as a gateway in the near future can be seen to multibillion-dollar industry.

For the various bioactive agents, microalgae have a great genetic potential. To produce these compounds this is proven as the ability of microalgae places these microorganisms in the biotechnological spotlight for applications and commercialization as in the pharmaceutical industry.

As the vitamin algae have also been examined and the vitamin precursor sources such as ascorbic acid, riboflavin, and tocopherol. For the various pharmaceuticals and nutraceuticals yet only, few micro-algal species have been examined. For the discovery of various new metabolites and high value compounds have large diversity of micro-algae makes it a potential candidate.

3.6.2. Antiviral Activity of Microalgal Compounds

To investigate the antiviral compounds, there is an urgent need that can be derived from the microalgae. For example, out of the 600 cultures of Cyanobacteria, and for the inhibition of the cellular infections different extracts were screened like HIV-1 and HSV-2, but it is being showed only the success rate 10%.

From the cyanobacteria named cyanovirin-N Patterson isolated a new compound, which was demonstrated as a powerful virucidal agent against HIV, by blocking the interaction of viral glycoprotein gp120 with CD4. On this compound there is a latent of developing a relevant vaginal microbicide against HIV based. Against various viruses, *Spirulina platensis* showed antiviral activity as Human cytomegalo viruses, Herpes simplex, and measles virus by blocking their entry due to the presence of Sulphur containing polysaccharide. Furthermore, by blocking the adsorption of virions against HSV-1, HSV-2 the enhanced antiviral activity due to Sulphur containing polysaccharides has also been shown by a red alga.

3.6.3. Anticancer Activity of Microalgal Compounds

From the microalgae, one of the compounds which are intended to maintain a focus and various bioactive compounds having anticancer activities. For antineoplastic activity, about 1000 extracts were screened from cyanobacteria and the activity isolated from the surroundings. due to the presence of inhibitors on differentiation of human leukemic cells, this activity was shown, and a success rate of 7% was recorded.

It has been seen that in the other reports, a study was conducted by using mechanism-based assays which also contain a various extracts like various enzymes such as protein kinase C and protein tyrosine kinase. By the process of apoptosis, it has been seen that there are many cyanobacteria that are promising producers of bioactive compounds and capable of destroying cancer cells, and by activation of signaling enzymes of protein kinase C family cell signaling will be affected.

Against the implanted lymphocytic leukemia and lung carcinoma, these compounds are more effective. With the higher anticancer activity, a metabolite named cryptophycin which was screened from Nostoc ATCC 53789, whereas a protein serine inhibitor named scytonemin isolated from Stigonema, may provide an admirable drug with antiproliferative and anti-inflammatory activities.

3.6.4. Cosmetics

In the skincare market, microalgal species like Arthrosporic and Chlorella are well established players, and some companies like LVMH, Paris, France and Daniel Jouvance, Carnac, France have landed in their own microalgal cultivation systems.

3.6.5. Pigments

In the microalgae, it has been seen that a large number of pigments associated with light occurrence which are being founded. Expect chlorophyll as primary photosynthetic compound, the important ones are phycobiliproteins and carotenoids. In the microalgae, a large number are being found. From the microalgae carotenoids extracted have various applications in market: β-carotene from Dunaliella as vitamin supplement in health foods, Lutein, zeaxanthin and canthaxanthin for pharmaceutical uses and chicken skin coloration and Astaxanthin for aquaculture to provide naturally red color for some fishes like salmon, extracted from Haematococcus. As the food and the cosmetic applications, the phycobiliproteins like phycocyanin and phycoerythrin, which are unique in algae are already in use.

For the human use, it has been seen that the antioxidant activity of carotenoids is of high importance. Due to their role as free radicals' scavengers, the antioxidants shows anti-cancer effects. Astaxanthin is of high potential among the different natural anti-oxidants.

Nowadays β-carotene in used in health foods due to its antioxidant effect and role as vitamin A precursor. Pigments which are being extracted from the microalgae are used as natural food colorants and cosmetic ingredients. It has been seen that there are a number of algae that contain an appreciable amount of carotene except β-carotene. to improve the health and fertility of grain fed cattle, there are β-carotene which is commonly known as a food colorant or color enhancer in fish flesh, yolk of eggs.

3.6.6. β-Carotene

As the natural food coloring agents' carotenoids are generally being employed with the additives for cattle feeds and in various cosmetics. There are various carotenoids such as β-carotene act as provitamin A, in terms of nutritional importance. Other than the synthetic compounds, there are many physical properties of natural β-carotene make it a superior candidate. β-Carotene is specifically a fat-soluble.

There are various investigations that are being conducted and found the effective nature of β-carotene for controlling cholesterol and preventing the risks of heart disease and recently National Cancer Institute declared β-carotene as anticarcinogenic. Hopefully, to boost the order of product in the commercial sector, such new studies make β-carotene as a potential candidate.

As the food colorants, the application of microalgae is restricted, because with heating of non-photo stability of food colorants and bleaching of color. The promising market for microalgae-derived food colorants is boundless in exception to all these disadvantages. A photosynthetic pigment microalga is being cultivated for β-carotene.

The uses of β-Carotene are: as orange dye and vitamin C supplement. There are many countries like Australia, Israel, the USA, and China, they have established various commercial plants for cultivating these microalgae. To the synthetic β-carotene, it has been seen that β-carotene derived and has more antioxidant property in comparison to synthetic because the latter is comprised of only trans-isomers while algae products have both cis and trans isomers.

3.6.7. Astaxanthin

From the microalgae, a different carotenoid is astaxanthin which has industrial applications. From the alga *Haematococcus pluvialis*, this pigment is a keto-carotenoid, mainly derived and produced during unfavorable conditions, when thin flagellated algae cell walls change to red thick-walled resting stages due to 4–5% dry weight.

As the food colorant this pigment is generally used as, feel additive for poultry industry and as feed additive for fishes like salmon, trout and shrimp. With the average price of US$2,500/kg the aquaculture market of this pigment is net US$200 million annually. In the form of encapsulated nutraceutical astaxanthin has been sold for human use.

This product has many health benefits about 100 times greater than tocopherol due to its antioxidant activity. Against the various diseases, various reports have shown the positive effects of astaxanthin example, cancer, metabolic disorders, diabetes, eyes diseases, and neurogenerative disorders.

3.6.8. Phycobiliproteins

In the different types of algae, the water-soluble accessory photosynthetic pigments found such as cyanobacteria, rhodophytes cryptophytes, glaucophytes and commonly known as phycobiliproteins. These are further divided into three categories: phycocyanin, allophycocyanin, phycoerythrin differing only in their spectral properties.

In the algae, the number of various elements like phycobiliproteins and it is being altered with environmental parameters like intensity of light and spectral quality of light. For example, from 11 to 12.7% dry weight, it has been seen that *Spirulina platensis* when grown at various light intensities, the amount of phycocyanin varies.

For the production of phycobiliproteins, Spirulina and Porphyridium are two common microalgae, which are commercially exploited. In food, these pigments have main potential as natural dyes, cosmetic products, and diagnostic tools in biomedical research as fluorescent markers.

For example, under a brand name pigment phycocyanin is advertised called Lina blue by Dainippon Ink and Chemicals for use in popsicles, candies, cold drinks, dairy products, and chewing gum. From US$ 3 to US$ 1500/mg the price of phycobiliproteins vary for some cross-linked pigments.

3.6.9. High-Value Molecules

For the production of high value compounds out of 30,000 known species of microalgae, only a few are harnessed such as pigments or proteins. Currently, as a hidden source of various new bioactive compounds microalgae represent itself.

To exploration of various novel bioactive compounds can be viewed as an endless ground due to their large biodiversity, which the studies are being related. With the percentage of polyunsaturated fatty acids (PUFA), the important sources of various bioactive lipid compounds are marine microalgae and are effective against numerous diseases.

The prevention of several diseases which include cardiovascular disorders, cancer, asthma, arthritis, kidney and skin disorders, depression and schizophrenia has also been shown by polyunsaturated fatty acids (PUFA) like linolenic acid, eicosatetraenoic acid (EPA), docosapentaenoic acid (DPA) and docosahexaenoic acid (DHA). Among various algae species, *Daniella* sp., *Chlorella* sp, and *Spirulina* sp. are major producers of high value compounds such as proteins, lipids, and pigments.

3.7. ROLE OF GENE TECHNOLOGIES IN BIODIESEL PRODUCTION

In solving the world energy crisis, genomics can play a key role. Scientists can develop a better approach of how to harness various renewable sources of energy through genomics such as cyanobacteria, microalgae, and lignocellulosic biomass. On a global scale, one day, the development of sustainable biofuel through genetic engineering of an enzyme will replace fossil fuels.

For the liquid fuel, cellulose biomass has the capability to meet the demand, but to-biofuel technologies for high scale deployment, process inefficiencies, and land-use requirements are the main obstacles. Therefore, the biomass can get breakdown by the genomics knowledge of microorganisms and the potential energy crops will be vital for improving the prospects of important cellulosic biofuel production.

The domestication of these plants the genomic information of bioenergy crops will also accelerate and even will improve their growth on poor agricultural lands to avoid competition with other food crops. In addition to it, for light and the growth rates competition such as cell wall chemistry, stem thickness, branching habit and it will maximize biomass yield per unit land area.

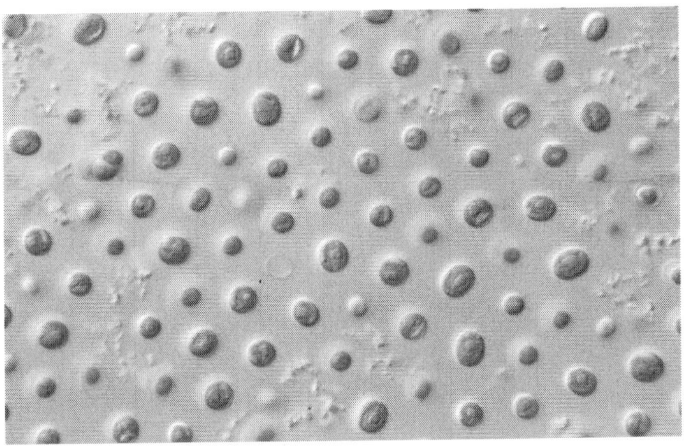

Figure 3.11: Role of gene technology in biofuel production.

Source: Image by Wikipedia.

Currently for increasing and modifying the accumulation of energy-rich compounds there will be intensive global researchers such as polysaccharides, hydrocarbons, alcohols, lipids, and other energy-rich chemical compounds in photosynthetic organisms, bacteria, and yeast *via* genetic engineering have been carried out with good results.

Though in photosynthetic microorganisms, many advances through genetic engineering have been achieved, because of their capability to secrete or store energy-rich hydrocarbons, superior growth rates, diverse metabolic abilities and high photosynthetic conversion efficiencies. To extend these advances to industrially relevant microorganisms' further studies are needed.

3.8. COMPARATIVE ANALYSIS OF ALGAL GROWTH AND THEIR OIL CONTENTS PRODUCTION

On the spirulina it is being worked, which is a blue-green alga. In the closed system they developed the cultivation method of this alga. For space they were developing this method to sue this algal species. For growing of algae homogenous, this development was beneficial to keep the conditions. To obtain the products of algae, this was a good step.

It has been seen that the system is being made in such a way the carbon dioxide and oxygen control was made beneficial. Furthermore, product isolation was made easy by the separation of algae and biomass. During the period of January to June 1998, Leghari carried out the ecological study of algal flora of the Jhelum River in Azad Kashmir.

From the seven groups of algae this include the research on 134 species, belonging to 68 genera. to assess the population of this algal flora as they play an important role in a balanced ecological system and this is the basic idea of the study. To dangerously high levels water pollution has increased the pollution of the flora, inducing oxygen-depletion in water, which unbalances the ecosystem as is hazardous for aquatic life.

As they are the primary producers' scarcity of such plants may also be dangerous, and their scarcity may unbalance the aquatic food chain. Therefore, a balanced population of the algal flora is necessary for a balanced ecosystem. The comparative ecological study of algal genera carried out by the Leghari and useful aquatic weeds of various localities.

Belonging to 5 classes 29 genera of algae, 47 aquatic weeds belonging to 21 families of 5 groups and 35 physicochemical parameters were recorded. In primary productivity and fish production, lakes are rich. To search the algal flora, the study revealed and aquatic weeds, water analysis to observe physicochemical properties, comparative study to explore species either rare or common to observe and are beneficial or toxic in use.

The limnological studies is being reviewed by Ali, keeping in view their strong impact on wetland's biodiversity. The results revealed that among total 103 species of planktons, 51 were found in Uchalli, 47 in Khabbaki and 39 in Jahalar Lake, and 31% species were phytoplankton, whereas

69% were zooplankton. Chisti (2008) revealed that because of the higher photosynthetic efficiency of algae, higher biomass productivities, a faster growth rate than higher plants highest CO_2 fixation and O_2 production, growing in liquid medium which can be handled easily, algae can be grown in variable climates and non-arable land including marginal areas unsuitable for agricultural purposes (e.g., desert and seashore lands), in non-potable water or even as a waste treatment purpose, use far less water than traditional crops and do not displace food crop cultures; their production is not seasonal and can be harvested daily.

The screening of microalgae is being investigated by the Gouveia, which was performed to choose the best one(s), in terms of quantity and quality as an oil source for biofuel production. N. It has been proved to be suitable as raw materials for biofuel production because of their high oil content. Under the nitrogen shortage when grown both microalgae showed a great increase (50%) in oil quantity.

Only from one species S if the purpose is to produce biodiesel. Obliquus presents the most adequate fatty acid profile, namely in terms of linolenic and other polyunsaturated fatty acids. Therefore, the microalgae *N. oleoabundans*, Nannochloropsis sp., & *D. tertiolecta* can also be used if associated with other micro algal oils and vegetable oils. Sharif et al., (2008) studied the common species Oedogonium and Spirogyra to compare the amount of biodiesel production.

However, biomass was higher in Spirogyra than Oedogonium sp. Sediments. Between the Spirogyra and Oedogonium ,species there was no difference of pH. From both the species and Oedogonium, these results indicate that biodiesel can be produced is a better source than Spirogyra species.

From a variety of wastewater sources, Park reported that algae could use growth nutrients such as nitrogen and phosphorous, therefore providing a sustainable bioremediation of these wastewater for economic benefits.

3.9. AQUATIC PLANT AZOLLA AS THE UNIVERSAL FEEDSTOCK FOR BIOFUEL PRODUCTION

To be ideal bioenergy crops a systematic analysis of terrestrial plants for their potential which led to a number of important selection criteria which include: chemical composition and high proportion of biofuel-producing molecules, high growth rates/biomass production, high harvest index/

rotation period, ability to grow on marginal lands and lack of competition with agricultural crops for arable lands, high freshwater use efficiency and low growing cost, low harvest cost and production of high value co-products. For the domestication of the next generation of energy crops, these criteria have triggered as an intensive search.

With their evolutional cyanobacterial symbiont, Azolla species are being together, which attracted the attention because of their potential to address all of the criteria mentioned above. Firstly, it has been seen that Azolla's biomass contains major types of energy molecules which include cellulose/hemicellulose, starch, and lipids resembling combinations of lignocellulosic, starch- and oil-producing terrestrial bioenergy crops and microalgae/cyanobacteria. Secondly, by doubling their biomass every 5–6 days and high annual productivity, they are one of the fastest-growing plants, being second after microalgae for biomass production rates. Similar to microalgae their biomass production is not seasonal. Thirdly, Azolla can be grown efficiently outside of their natural habitat using wastewater as their main source of nutrients. Bioremediation of wastewaters can lead to absorption of up to 2.6 and 0.434 t/ha-year of total N and P, respectively. It has been seen that Azolla can also grow efficiently in N-depleted media assimilating up to 0.4 t N/ha-year nitrogen from the atmosphere through symbiosis with *A. azollae.*

Fourthly, they are easy to harvest. And finally, Azolla species have been used for decades as nitrogen bio-fertilizers and livestock/fish feed (due to their high protein and carbohydrate levels). The lack of need to use synthetic nitrogen-based fertilizers for growing Azolla makes a substantial positive impact on the reduction of nitrous oxide (N_2O) generation, the impact of which on warming the atmosphere is almost 300 times that of carbon dioxide.

3.9.1. Thermochemical Technologies

For the production of bio-oil, bio-char and biogas Azolla biomass was used as feedstock by using the two thermochemical technologies, pyrolysis and the HTL. The theoretical production of pyrolysis- and HTL-based bio-oils valued at up to 13.2 and 20.2 t/ha-year, as it is given by the high annual productivity of Azolla, respectively.

It has been said that both bio-oils contain a range of petrochemicals that can be directly used as glycerin-free components of biodiesel or can be upgraded to biofuels using some well-established technologies. Biosolids can be produced by both thermochemical technologies such as biochar and ash (up to 48 t/ha-year for both), which can be used to improve soil quality by increasing their nutrient- and moisture-holding capacity, support microbial communities and promote the root activity.

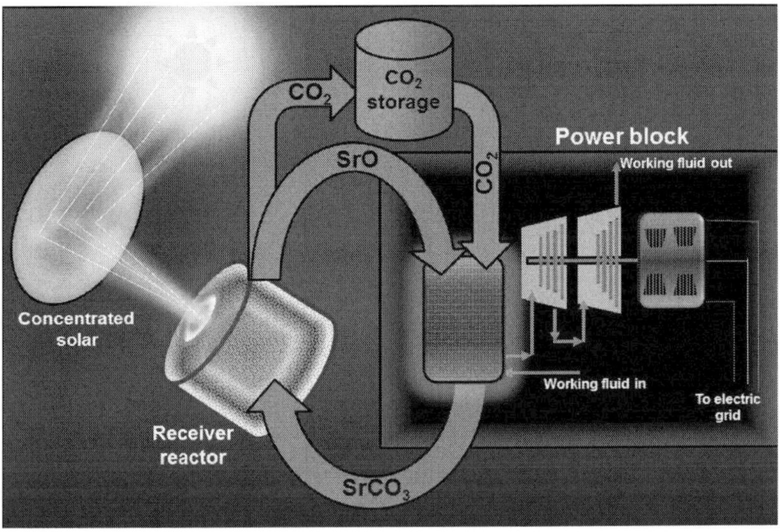

Figure 3.12: An illustration of thermochemical energy plant.

Source: Image by Flickr.

Alkali elements are the one ash from the Azolla contains as well as microelements such as Fe, Mn and Cu making it a good candidate for use as fertilizer.

3.9.2. Biodiesel Production

The annual theoretical yield of the crude lipids from A which is being grown on a wastewater. It can be estimated as 1.68 t/ha-year. If the plant will grow under natural conditions with productivity around 100 t dw/ha-year this amount can be increased to 8 t/ha-year. From the soybean, sunflower, rapeseed and oil palm this oil productivity is significantly higher, but lower than the theoretical yield from microalgae. Azollae contains a range of C16:0, C18:2 and C18:3 fatty acids, their conversion to methyl esters means that the resultant biodiesel meets the crucial requirements of fuel density, cetane number and iodine value for biodiesel set by the EN14214 standard.

3.9.3. Ethanol Production

From the theoretical ethanol production, folliculinids is 9.3 t/ha-year (11.7 × 103 L/ha/year, based on specific volume of ethanol = 0.789 g/mL) which is lower than ethanol production from sugarcane (25 × 103 L/ha-year), close to that from corn stover (13.31 × 103 L/ha-year), but higher than from miscanthus (2.3 × 103 L/ha-year), willow (0.3 × 103 L/ha-year) and poplar (1.3 × 103 L/ha-year).

3.9.4. Bio-Hydrogen Production

It has been seen that for making the Azolla a competitive feedstock for hydrogen production compared with other bioenergy crops, the high biomass production yield, chemical composition (C/N ratio) and high bio-hydrogen yield of 2.2 mol/mol substrate.

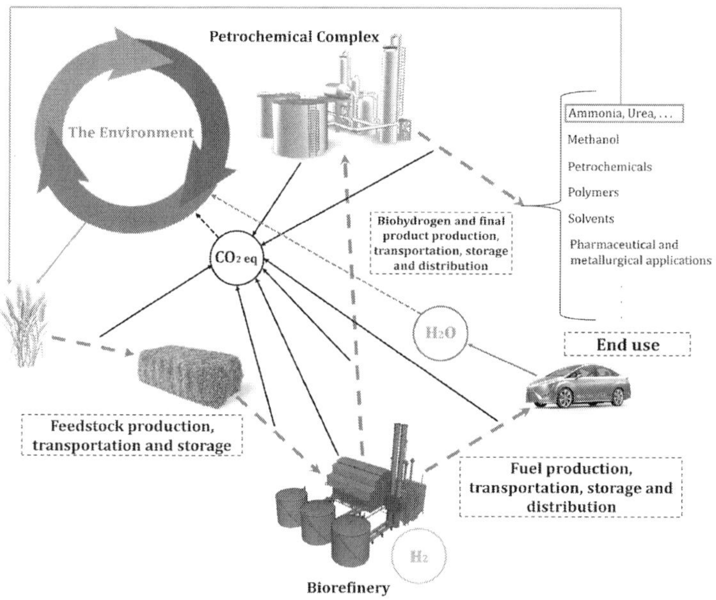

Figure 3.13: An illustration of bio-hydrogen production and its use.

Source: Image by Wikimedia Commons.

3.10. CONCLUSION

Overall, it has been seen that there are some advances that have been made in the biofuel production. As, with the adaptation of new technology and also by confirming that how challenges should be faced regarding the land use, water use and the nutrient challenge.

By analyzing all the algae-based nonenergy options, the production is become more convenient. And the role of Gene technologies also helped in solving the economic crisis for all the developing countries. Advances which are there is also because of the comparison which is being made between the algal growth and their oil content production.

REFERENCES

1. Food and Agriculture Organization of the United Nations, (2010). Algae-based biofuels: applications and co-products. [ebook] Food and Agriculture Organization of the United Nations. Available at: http://www.fao.org/3/i1704e/i1704e.pdf (accessed on 31 July 2021).
2. Khan, S., Siddique, R., Sajjad, W., Nabi, G., Hayat, K., Duan, P., & Yao, L., (2017). Biodiesel Production from Algae to Overcome the Energy Crisis. *HAYATI Journal of Biosciences,* [online] *24*(4), 163–167. Available at: https://www.sciencedirect.com/science/article/pii/S1978301917300190 (accessed on 31 July 2021).
3. Michael Hannon, Javier Gimpel, Miller Tran, Beth Rasala, & Stephen Mayfield, (2010). Biofuels from algae: challenges and potential. [online] PubMed Central (PMC). Available at: https://www.ncbi.nlm.nih.gov/pmc/articles/PMC3152439/ (accessed on 31 July 2021).
4. Miranda, A., Biswas, B., Ramkumar, N., Singh, R., Kumar, J., James, A., Roddick, F., Lal, B., Subudhi, S., Bhaskar, T., & Mouradov, A., (2016). Aquatic plant Azolla as the universal feedstock for biofuel production. *Biotechnology for Biofuels*, [online] *9*(1). Available at: https://link.springer.com/article/10.1186/s13068-016-0628-5 (accessed on 31 July 2021).
5. Sharma, P., & Sharma, N., (2017). Industrial and Biotechnological Applications of Algae: A Review. *Journal of Advances in Plant Biology,* [online] *1*(1), 1–25. Available at: https://openaccesspub.org/japb/article/530 (accessed on 31 July 2021).

Chapter 4

Marine Microalgae in Food and Health Industry

CONTENTS

This chapter basically explains the marine microalgae in food and the health industry. It further highlights the current uses of the microalgae as food. This chapter gives details about the role of microalgae within the food and health industry as these two are the currently most demanded industry.

There are various uses of microalgae as the nutraceuticals and food, which have been highlighted in the below-mentioned context. This chapter also highlights the role of the microalgae in the micro and macronutrients and why they are useful for the diet and this context later describe the use of the microalgae and the use and the importance of the microalgae in the sustainable production.

There are many more aspects of the microalgae which have been clearly explained in this chapter. This includes the toxicological aspects, the nucleic acids the algal toxins. The microalgae play a vital role in the health and applications, and it has highlighted and along with the important details. The detail information regarding the use of microalgae and cyanobacteria for health has been explained well.

There are various properties of the microalgae which are been described in this chapter; basically, the microalgal compounds possess the antihypertensive properties, which are important to understand the concept clearly.

4.1. INTRODUCTION

To provide enough food for the population of the entire planet is currently one of the major challenges in the present scenario. It has been basically estimated that sufficient water and arable land is not available to support such demands and this estimation is made on the basis of the predictions of the world population will have increased by another 2 billion by 2050 as per the prediction there is an explicit need of the food and demand for the upcoming generation. Further, over the traditional land and the plants, there are various advantages of the microalgae. Some of the advantages are included below:

1. With the help of using photosynthesis, efficient conversion of carbon dioxide (CO_2).
2. Within in a short span of 3 to 5 days, high growth productivity could be seen along with a very short harvesting cycle when it is clearly compared with the other feedstocks span of 8 to 10 days, and these are also being harvested mainly once or twice a year.

3. Further, using saltwater that may be brackish waters of saline, they can be easily grown also there being able to grow in non-agricultural land (e.g., desert, seashore or semiarid regions) and also in the marginal wastewater (domestic/agricultural/municipal/industrial wastewaters).
4. Further within the biorefineries, along with the compatibility and also with the integrated production of biofuels is being obtained.

The above-mentioned are the clear advantages of the microalgae and these characteristics also make them suitable for the supply for the commodities. Also, they provide opportunities basically for both non-food products and food products.

Figure 4.1: Basic significance of marine microalgae in food and health industry.

Further for accumulations of macromolecules which are namely are carbohydrates, proteins, lipids, etc., and they are basically accumulated by the efficient use of the solar energy. These microalgae could be both prokaryotic cyanobacteria and eukaryotic microalgae.

Due to their ability to synthesize valuable products mainly like caretonoids, abstract microalgae which also includes cyanobacteria are the promising organisms which are in turn sustainable products for the use as feedstocks for the food, feed, chemicals and biofuels and in addition to this the microalgae are considered as the perfect candidates for contemporary nutraceutical or as the functional food.

Further with medium to high value market price such as β-1,3-glucan polysaccharide, phycobilin pigments, carotenoids and also the polyunsaturated fatty acids, these are basically commercialized within the food industry. These are moreover commercialized as the functional foods, dietary supplement and also as the flavorants and cosmeceuticals in the pharmaceutical and the chemical industries, as nutraceutical compounds within the therapeutic field.

Acids such as eicosapentaenoic, docosahexaenoic and arachidonic acids are fatty acids, particularly omega-3 and omega-6, and microalgae are exceptionally resembling their producers. These above-mentioned fatty acids are basically linked to the several human health benefits.

Further, with a particular focus on the food science and the technology applications, there are some of the main existing high value products which have been derived by the means of the microalgae. And further the entire chapter reviews the functions or the impacts of the microalgae basically in the food and the health industry.

There are various details mentioned in regard to the nutritive importance of the selected to the constituents, and this chapter also further describes the functions of the microalgae in the different aspects of the food and health and how it has been tremendously being acted in the current scenario.

Further, it has been observed that mainly to ensure the food safety in the use of the algal biomass are presented and also there are various nutritional quality standards and the legislative provisions regarding this. Dr. James A. Kenar, Editor-in-Chief JAOCS. Ângelo Paggi Matos is the one from whose invitation some of this context is being taken as it provides tremendous information.

Mainly for the human nutrition, there are some of the essential components which are namely the sugar, essential and the non-essential amino acids, long-chain fatty acids and also the minerals and the vitamins.

Further, there are some of the most important biotechnologically relevant microalgae which are being widely commercialized as the nutritional supplements for the humans and as well as for the animal feeds additives.

The above-mentioned supplements are namely the green algae further classified as the Chlorophyceae, *Chlorella vulgaris*, *Haematococcus pluvialis*, and *Dunaliella salina* and the cyanobacterium (also known as blue-green algae), and *Arthrospira platensis* (Spirulina).

Also, there are various cosmetic purpose which can basically be fulfilled by the means of the algae as it acts a chemical platform for all these mentioned purposes they can be for example as coloring, pigment of skin creams also for the therapeutic applications as well as for the pharmaceuticals. Further, as powerful pigment factories, some of them are being highlighted they are namely the Thchlorophyceae Haematococcus and Dunaliellaa and also with an outstanding capacity for the protein accumulation is the *Spirulina platensis*.

It has been lately observed that in the form of the triacylglycerols (TAG), such as omega-3 (ω-3): α-linolenic acid (ALA), eicosapentaenoic acid (EPA), and docosahexaenoic acid (DHA) and omega-6 (ω-6): linoleic acid (LA), γ-linolenic acid (GLA) and arachidonic acid (AA) are basically the long-chain fatty acids (LC-PUFAs) which are basically accumulated in the substantial amounts.

These long-chain fatty acids are also linked to the health effects of the humans which includes to the anti-inflammatory effects, macular degeneration, rheumatoid arthritis, and also the mitigation effects of the hypertension.

There are some of the products are which are commercialized as the single-cell oil (SCO) containing LC-PUFAs, include *C. vulgaris* (ALA-producer), *S. platensis* (GLA-producer), *Isochrysis galbana*, *Nannochloropsis oculata* and *Phaedo-dactylum tricornutum* (EPA-producers), *Crypthecodinium cohnii* and *Schizochytrium limacinum* (DHA-producers) and *Porphyridium cruentum* (AA-producer) and these are basically the edible oil which are mainly reported by the algae.

Further, mainly to provide the superior nutritional supplement through the consumptions of LC-PUFAs, there are SCO which have been attracted attention of the several of their commercial development in order to reach the explicit demand of the market.

Also, a schematic overview of potential microalgal cell factory which are namely the lipids, protein, polysaccharides and bioactive compounds and these are basically correlated to its areas of final applications and these are mainly the food technology and the fine chemicals, pharmaceutical, nutritional purposes, and agriculture and all of these above-mentioned are basically derived compounds and they are mainly the derived compounds of the products.

In several papers, there are various products that have been mentioned and can be reviewed. Further, this context basically allows or infers the aim. And mainly the aim of this chapter is to provide the up to date and comprehensive information regarding the uses of the microalgae in the food and the health industry. Also, with the particular focus on the use of these high valued products in the food science and the technology also provides the information regarding the commercial high value products which are being available or synthesized by the means of the microalgae are also being mentioned in the different chapters in this book.

Also, the nutritional quality standards of algae for human consumption are also discussed in addition to the toxicological aspects in terms of the heavy metals, algal toxins, and heavy metals.

4.2. CURRENT USES OF MICROALGAE AS FOOD

Further it has been observed that microalgae are not being the new available food but they have been a part of the human life since centuries. So, basically the above line infers that the microalgae as the food are being traditionally used. There are various examples such as Spirulina by certain tribes in Africa and Nostoc has been used in Asia.

Some of the interesting and the explicit accounts of the documented culture methods are being used in the cyanobacteria and the uses of the microalgae, and all of this is being reported by Gantar and Svirčev (2008). And all of this has been done by the indigenous populations in Mexico and other native populations.

Further in the health food market, the microalgae biomass has been used almost exclusively since the past decades. Also to the manufacturers of the powders, tablets and the capsules of the microalgae production are solely dedicated mainly by the annual biomass production, which is over 75% (Pulz and Gross, 2004; Hudson, 2008).

Figure 4.2: Current uses of microalgae as food.

Further in the food-based products, all the microalgae biomass and much more could be incorporated easily. Also, in various ways which are previously being unrealized and this industry also allows to grow and diversify in ways that were previously unrealized.

It has been noted that it is necessary to diversify to the other end products which are needed to ensure the continuing development of the microalgal tablets. And this has been be mentioned by Liang and others (2004).

Further, it has been observed that there all the microalgae biomass and much more could be incorporated within the food-based products. This industry basically allows to diversify and in the previously unrealized ways.

4.3. MICRO- AND MACRONUTRIENTS FOUND IN MICROALGAE

(Tokusoglu and Ünal, 2003; Becker, 2004; Liang et al., 2004; Shimamatsu, 2004; Spolaore and others, 2006; Rodriguez-Garcia and Guil-Guerrero, 2008). Further, for the different strains of microalgae under the different growth conditions there should surely many analyses of gross chemical compositions should be published. Moreover, protein content such as the efficiency ratio, the apparent biological value, and the true digestibility value of protein content are basically the new evaluations of the microalgal protein content mainly (Becker, 2004).

It has been lately observed that in the proportion with availability of amino acids in their protein profile and also in all these tests the microalgae have been compared to the favorably with the reference. Further, mainly for the human consumption, the microalgal species are being commonly utilized.

4.4. MICROALGAE: USES AS THE NUTRACEUTICALS AND FOOD

It has been observed that the microalgae have a wide range of industrial applications, and these industries are mainly the wastewater purification, pharmaceuticals formulations, in the food industries.

Further there is a wide aspect of the microalgae, and they can also be used in the preparation of the biofilms, dyes and feed for aquaculture, anti-inflammatory objects, cosmetics, food and fodder additives, health food for humans, high valued food and also for the dyes and the feed for aquaculture.

There are namely those species which are being widely used, and they are Cyanophyceae (blue-green algae), Chlorophyceae (green algae), Bacillariophyceae (which also included diatoms) also various products, microalgal species, and their applications.

4.5. MARINE MICROALGAE FOR SUSTAINABLE PRODUCTION OF FOOD AND FUEL (2018–2019)

The three greatest challenges which are being faced in the current scenario are basically the food security, energy, and the climate.

Further to be a promising, environmentally favorable approach for the society to meet its climate goals which includes the sustainably co-producing liquid hydrocarbon fuels and the protein by the means of the large-scale industrial cultivation of the marine microalgae which basically shows the above-mentioned goals as determined.

Further, it is being shown that the sustainable co-production of food and fuel is approaching economic feasibility mainly depicted by the current models, based on the limited available empirical data.

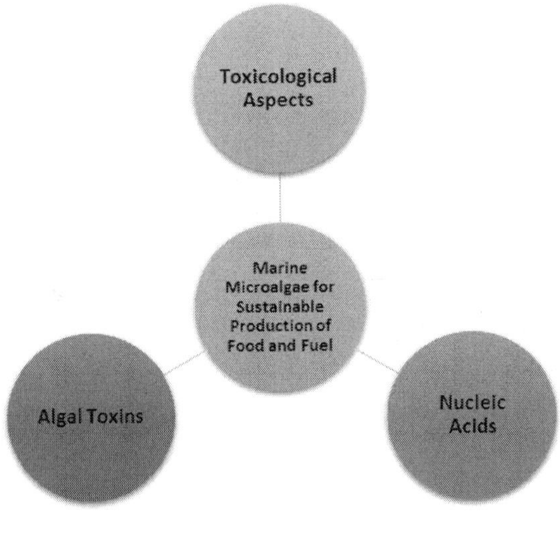

Figure 4.3: Marine microalgae for sustainable production of food and fuel.

A dramatic help to the advance the above-mentioned predictions when they are combined with a forward looking economic and life cycle analysis could mainly be provided by the emerging data from the Duke, which explicitly leads to the marine algae industrialization.

4.5.1. Toxicological Aspects

Further, there is basically a series of toxicological tests are there which prove their harmlessness as microalgae which are considered as the unconventional food. Lately, there are various international organizations, which have given several recommendations and toxicological evaluations they have been published by these organizations.

There are specific recommendations that are further required for the analyses, and these provide the assumed additional national regulations.

Finally, in the two aspects mainly there is an explicit need to be mainly analyzed in their terms basically as the part of the toxicological characterization and these compounds are being synthesized by the alga itself and they basically contains the biogenic toxins or the compounds which accumulate from the environment these are the non-biogenic toxins.

Non biogenic products which basically comprise the environmental contaminants such as the heavy metals and the biogenic toxins mainly include the nucleic acids and the algal toxins.

4.5.2. Nucleic Acids

The rich sources of the purines are basically including the RNA and DNA and it has been observed that the nucleic acids they are the biopolymers which are essential to all the forms of the life. The gout disease is basically caused in general by the association between the purine rich diet and which possibly increased the plasma urate concentration along with the association of the purine rich.

Further for the gout patients, there are certain recommendations mentioned which are mainly due to the uric acid as it is the end product of purine degradation and the avoidance of purine rich foods and these foods are highly recommended to these gout patients.

These are also sometimes being considered as the limitation for the food application, and it has been observed that the microorganisms are the sources of the single-cell protein and they in turn may contain the purines.

Moreover, there are various components which are being contained in it these are namely the algae which contain the 4–6% of the nucleic acid, bacteria contain up to 20% of nucleic acids in the dry matter while also in case of the yeast there is mainly 8–12%.

Mainly by the Protein Advisory United Nations (nutrition bulletin) basically highly recommends a maximum daily intake of 4.0g/day nucleic acids for the unconventional food source and this all because of the possible health hazard.

Further, mainly over the bacteria and the fungi sources due to their nucleic acids content as mentioned earlier, the single-cell protein from the algae sources are basically preferred over them. The safe level of algal biomass consumption is 20 g of algae per day or 0.3 g of algae per kg of bodyweight, mainly according to the Becker.

4.5.3. Algal Toxins

In both the seawater and the freshwater basically, the algal toxins are produced by the various algae are found mainly. As these do not contain any sort of smell or any kind of taste so there is clearly no elimination method required for their removal, which may be the cooking or else the freezing method, which further reduces the efforts.

Further, it has been observed that in the several regions of the globe, this can lead to the poisoning of humans as the algal toxins mainly occur from the blooms of the algae with the considerable frequency sometimes which are unpredictable.

It shows its effects on the other aspects such as fish die of which states having to close fisheries. Further, it has been noted that the are various impact on the human health mainly by the consumption of the contaminated fish oysters, scallops, clams, shellfish, and mussels which mainly filter these toxins from the water accumulating levels to the consumers also including the human beings and these are certain microalgae which shows the above characteristics and they are namely Mycrocystis sp., Dynophysis sp., and Pseudonitzschia) produce potent toxins (saxitoxins, brevetoxin, domoic, and okadaic acids.

There are basically various widespread different poisoning syndromes which are basically caused by the means of the algal toxins and these diseases are listed below:

1. paralytic shellfish poisoning (PSP);
2. neurotoxic shellfish poisoning (NSP);
3. ciguatera fish poisoning (CFP);
4. amnesic shellfish poisoning (ASP);
5. diarrhoeic shellfish poisoning (DSP).

These are mainly the diseases that have been generally shown to harm the health of the human beings. Furthermore, there are various species of the algae which are commercially being used in the food supplements and in the aquaculture and the advantages of these species are that they basically do not harm the health mainly they do not produce toxins.

These species are namely the Spirulina (Arthrospira), *I. galbana*, *H. pluvialis*, *C. vulgaris*, *D. salina*, *P. tricornutum*, and *P. cruentum*.

There are large differences which are mainly being observed even within the same species as they mainly exist between the toxic and the nontoxic algae. Mainly for instance, it has been observed that the dinoflagellate strain *C. cohnii* has a GRAS (Generally Recognized as Safe) status by FDA for ω-3 DHA human food consumption also this dinoflagellates and diatoms are best known for their production of toxins.

These toxins can affect humans, and this explicitly proves to be a good example to understand the above-mentioned texts. Eventually, the applied for food or feed applications as it is very important to know the safety of the algae at strain level which in turn leads to the above-mentioned statement.

4.6. USE OF MICROALGAE AND CYANOBACTERIA FOR HEALTH APPLICATIONS

This research has been mainly extended in to the microalgae as it is easier to cultivate and this is inferred by that the principal research into the bioactive components from the environment and their health applications has basically stemmed from studying macroalgae.

Further, there are various ailments which need to be treated, and they include diabetes, high blood pressure, and the other disorders which are mainly related to the metabolism and the immune system. And this interest in harvesting bioactive components from the microalgae and the cyanobacteria to be used in the health applications has been basically increased in line with research efforts are being quest for the molecules to treat ailments.

Further for the pharmacological mainly to harvest the active ingredients for them one of the species of the microalgae has been studied mainly. Further, to obtain astaxanthin, is considered as a super antioxidant also in the freshwater *Haematococcus pluvialis* is the species that is being cultivated and exploited industrially.

Dunaliella tertiolecta is further another widely cultivated microalga and this species of the microalgae is highly tolerant to saline environments. It has been lately noted that this species of the microalgae is also used to harvest the ß-carotenes.

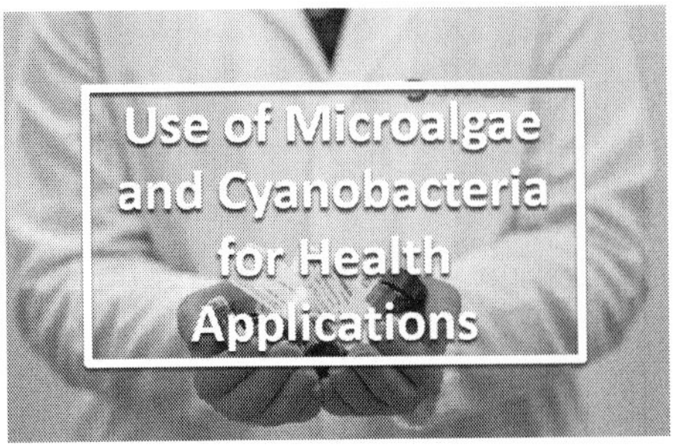

Figure 4.4: Use of microalgae and cyanobacteria for health applications.

Further too mainly produce the pigments, named as violaxanthin, which contains various properties, and the most important among them is the anticarcinogenic properties. Further, to obtain the phycoerythrobilin and the phycocyanobilins, proteins another species of the genus porphyridium are basically used to obtain them.

And also, been noted that the proteins colorings which are basically considered to be an alternative to synthetic coloring are mainly due to their marginal toxicity present in them and they also clearly used for harvesting the polysaccharides and they in turn possess the antiviral properties.

They are also been attributed with the anticarcinogenic properties likewise in some of the species from the chlorella genus (mainly *Chlorella vulgaris*) have been mainly used to obtain the chlorophyll for use as a pigment.

Basically, an active immune stimulator which rounds up as the free radicals and which also reduces the lipids in the blood are never the less most important substance in Chlorella is ß-1,3-glucan. There are some of the complex polysaccharides which basically possess the immune-stimulating properties are namely the species which are *Chlorella pyrenoidosa* and *Chlorella elipsoide*.

Further mainly due to the presence of the polysaccharides there are various properties which are mainly being possessed by the Chlorella and these properties are the anti-inflammatory and the analgesic abilities and this is being mentioned by the means of the other studies on this species.

Further due to the fact that there are many species which are created under the flexible cultivation conditions and the harvesting polysaccharides from the microalgae and cyanobacteria in particular the sulfated polysaccharides are the ones and also these species are helping out in attracting the growing interest.

They are further being used as the joint lubricants as these polysaccharides possess various properties, and they are also being demonstrated as the anti-inflammatory, immune-modulating, and the antiviral properties.

Further into the above-mentioned *Porphyridum cruetum* and *Porphyridium purpureum* also *Tetraselmis* sps., Isochrysis sps. The most popular species and the genera are these and in addition to these, all species are of chlorella among the others.

The production of lipids such as the triacylglycerol (TAG), is basically another highly interesting aspect in this.

They basically also possess the special interest for the human consumption as it contains the composition of fatty acids similar to many vegetables oils which also includes some of the high value such as the eicosapentaenoic acid (EPA) and docosahexaenoic acid (DHA).

Nannochloropsis gaditana, *Scenedesmus obliquus*, and the diatom *Thalassiosira pseudonana* are another species of the microalgae which basically helps to produce the TAG's. *Odontella aurita* is one of the species which is basically rich in the EPA and the other bioactive molecules and studies are mainly carries out on a marine diatom.

It has been shown that it is beneficial effect on reducing the metabolic syndrome risks factors which includes (hyperlipidemia, platelet aggregation, and oxidative stress and there are various effects are being encountered and they also help to the synergy between aggregation and the oxidative stress.

Phaeodactylum tricornutum, *Monodus subterraneus*, *P. cruentum*, *Chaetoceros calcitrans*, Nannochloropsis, *Crypthecodinium cohnii*, *Isochrysis galbana*, and *Pavlova salina* are some of the many other species of the microalgae which produce s the above-mentioned polyunsaturated fatty acids (PUFAs).

These are clearly used in the cosmetics. It has been lately suggested that the microalgae can be mainly utilized as an alternative and the sustainable source of the PUFAs for their use in the nutraceuticals and they possess these properties mainly due to this richness in the unsaturated fatty acids, and they only resemble these properties.

The entire species is marketed basically as the food supplement in the tablet, capsule, or powder form as the microalgae and the cyanobacteria are increasing tremendously use in the food industry. Further, it has been observed that only a handful of species are basically recognized as the ingredients for the use in the food industry, and the entire species is being marketed as a food supplement as there only a few of the species which are considered as the thousands that are known regarding these processes.

The above-mentioned species are the ones which are mainly being the proteins, peptides and amino acids from the *Chlorella vulgaris* and Spirulina.

There are some of the species which are being used as the lipids and the fatty acids Haematococcus and Spirulina, the ß-carotenes from *Dunaliella salina*, and the phycocyanobilins from Spirulina as antioxidants also there are furthermore such species.

Though there some of the uses yet to be developed but another nutritional aspect of the interest is the potential use of the polysaccharides as a source of fiber and also a prebiotic.

4.7. MICROALGAL COMPOUNDS WITH ANTIHYPERTENSIVE PROPERTIES

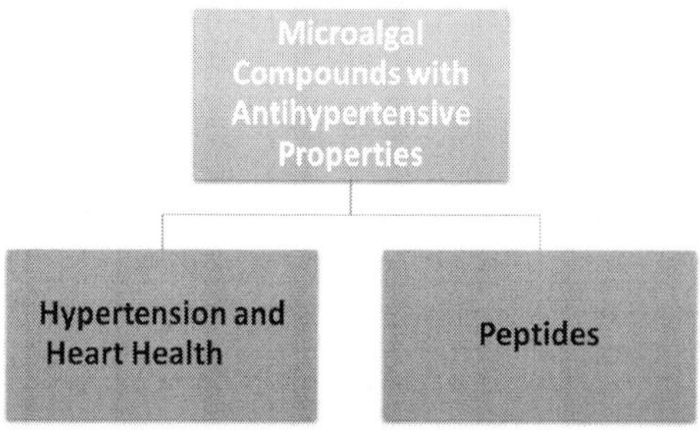

Figure 4.5: Microalgal compounds with antihypertensive properties.

4.7.1. Hypertension and Heart Health

One of the major health concerns worldwide in the current scenario is the hypertension. The risk factors of the chronic ailments which including the cardiovascular disease, stroke, renal disease and the diabetes and some of the main cause of the death are being recognized. Further, a multitude of complex systems which are mainly governed by the ACE in the rennin angiotensin aldosterone system are basically being recognized, and these are due to the blood pressures in the body.

It has been observed that the Angiotensin II Type 1 Receptor in the vascular smooth muscle cells to induce a potent hypertensive effect and ACE catalyzes the conversion of inactive Angiotensin-I (Ag I) to Angiotensin-II

(Ag II), a potent vasoconstrictor peptide that binds directly to the above-mentioned receptors.

Hypotension, angioedema, coughing, dizziness, headaches, nausea, and renal damage are associated with the unpleasant or intolerant side effects and for them, their specific treatments with ACE inhibitors. Thus, there is great interest in the use of antihypertensive compounds from natural resources. A major burden on the global health care costs and synthetic antihypersentitive drugs are represented.

4.7.2. Peptides

Bioactive peptides that are derived from the food have shown their antihypertensive characteristics with the help of the inhibition of enzymes that take participation in controlling the mammalian blood pressure, for example, ACE and renin.

Moreover, several numbers of various peptides have shown the capability in order to improve the endothelial nitric oxide synthase pathway, leading in direction towards the increment of the levels of nitric oxide within the vascular walls, in this way, they promote vasodilation.

Apart from this, the other peptides were reported to block the interactions between the Ag II, a vasoconstrictor, and angiotensin receptors, that plays an important role in decreasing the blood pressure. Nevertheless, ACE-inhibitory activity is the most-studied metabolic pathway for the development of the antihypertensive peptides.

Several numbers of sources of plant and animal proteins, primarily from the dairy products and fish, have been used in order to produce the ACE-inhibitory peptides but also, the application of the microalgal proteins has been very restricted.

Till the present interval of time, *Chlorella vulgaris*, *C. ellipsiodea*, *Arthrospira platensis*, and *Nannochloropsis oculata* have been utilized in order to produce ACE-inhibitory peptides. Their efficiency in decreasing the blood pressure has been showed a broad use in both *in vitro* studies as well as animal studies.

A purified peptide which is isolated from the pepsin hydrolysate of algae protein waste from *Chlorella vulgaris*, with the amino acid sequence Val – Glu – Cys – Tyr – Gly – Pro – Asn – Arg – Pro – Gln – Phe, was found to possess ACE-inhibitory activity with an $IC_{50} = 29.6$ **μM**.

Two purified peptides, Gly – Met – Asn – Asn – Leu – Thr – Pro and Leu – Glu – Gln, from the cultured marine microalga, *Nannochloropsis oculata*, reported ACE inhibitory activity with IC $_{50}$ values of 123 **μM** and 173 **μM**, respectively.

In the same way, a potential ACE inhibitory peptide from the marine *Chlorella ellipsoidea* which is made up with the help of only four amino acids, that are, Val – Glu – Gly – Tyr, was shown to have ACE inhibitory activity (whose IC $_{50}$ value = 128.4 **μM**) along with an antihypertensive effect in spontaneously hypertensive rats (SHR) when administrated orally.

The power of the antihypertensive effect is strongly impacted with the help of the structure and configuration of the bioactive peptides. The outcomes of analytical and chemometric experiments presented that ACE-inhibitory peptides preferred to bind to C-terminal or N-terminal catalytic sites of ACE.

In addition to this, it was recommended that substrates that are consisting of amino acids with cyclic or aromatic rings (e.g., Pro, Phe, Tyr, Trp) at their C-terminus and hydrophobic amino acids, more particularly those which are having aliphatic chains (e.g., Gly, Ile, Leu, Val), at their N-terminus were thought to improve the ACE-inhibition.

In spite of the evidence that displays the significance of the C-terminal domain in the form of blood pressure regulation, the N-terminal domain was considered as the most vital aspect with respect to the ACE inhibition. The advantages are comprised of mainly of higher blockage of Angiotensin II production and better prevention of organ damage.

4.8. ANTIVIRAL PROPERTIES OF MARINE ALGAE

Vaccines play a very important and effective role in the regulation of several numbers of various viral diseases, but still, there are few diseases that are not regulated with the help of the vaccination. Some synthetic antiviral compounds were created for the treatment of active herpetic infections, that were not that much successful and efficient for the treatment of latent infections.

It stated on severe side effects as well as the development of some resistant mutations of this virus, more particularly, through the course of long-term medication with antiviral drugs. And the reason behind this is the toxicity of several of the earlier antiviral agents. The idea of antiviral compounds with pharmaceutical value could not be accepted easily. Few plants and algae extracts were tested on various viruses that are consisting of the herpes viruses. In some of these experiments, various kinds of species of brown algae were tested for their antiviral activity.

With respect to the test report, it has been noted that these sorts of species are consisting of antiviral property. The antiviral agent must have an extensive spectrum of activity, provide assistance in prevent the virus entirely, to have favorable pharmacodynamics properties and not be immunosuppressive, which is important, because there should be no suppression of the normal immune processes.

The detection of the role of the interferon in the form of cellular antiviral systems and the elucidation of differences between normal cellular metabolism and viral replication have led to a renewed interest in antiviral chemotherapy.

4.9. ALGAL METABOLITES IN FOOD

In the present interval of time, all across the globe, the primary focus of the people is on their diet. The diet of highly caloric and in combination with the modern style of life will take into the direction towards the issues that are related to their health, for example, obesity, heart diseases, diabetes, and so on.

In this way, food products are required to be focused in order to promote health with the help of enriching the diet with vitamins, minerals, along with healthy nutrients, and as per the consumers that are using the natural forms of ingredients in spite of the synthetically ones have become very crucial.

Figure 4.6: Algal metabolites in food.

According to the reports, that have recommended that there is considerable enhancement in the state of the health not only to algal proteins but also to the therapeutic factors. In coastal areas of all continents, seaweeds are utilized in human and animal nutrition, in such a way that they are extensively cultivated algal crops.

In the present interval of time, consumers are fascinated towards species like Porphyra sp., Chondrus crispus, *Himanthalia elongata* and *Undaria pinnatifida*. In addition to this, they are also attracted towards the food industry, and the reason behind this is their low content in calories and high content in vitamins, minerals, and dietetic fiber.

Microalgal biomass is available in the form of powder, tablets, capsules, liquids and, also, the microalgal biomass can be combined into various kinds of food products and the most significant in human nutrition are Spirulina and Chlorella genera.

4.10. HEALTH BENEFITS OF MICROALGAE IN PREVENTING CARDIOVASCULAR DISEASES

The health benefits of microalgae are being more and more familiar and valued within the time period of last three to four decades as the introduction of probiotic nutritional supplements. Health benefits of food ingredients as well as the dietary supplements are attributed more importantly to the long-chain polyunsaturated fatty acids (LC-PUFA).

Several numbers of various microalgae, as highlighted in the previous section, they have the capability in order to synthesize long-chain polyunsaturated fatty acids with particular interest, more particularly, the omega-3 and omega-6 series, for example, EPA, DHA and ARA (Pulz and Gross, 2004).

Freshwater microalgae as well as the marine microalgae have been found to have good effects on and advantageous to cardiovascular diseases, inflammatory diseases, cancer along with the viral infections. This section of the chapter is more focused on the role of few microalgae that provide assistance in the prevention of cardiovascular diseases.

The microalga or more particularly cyanobacteria Athrospira sp. has been revealed to elevate the plasminogen activating factor that is present in the endothelial cells with a positive influence on the cardiovascular disease prevention.

This cyanobacterium also plays an important role in order to improve the immune system with a prevention in both viral infections as well as cancer infections, in addition to this, cyanobacteria also provide assistance in order to elevate the gastrointestinal microorganism flora (Barrow and Shahidi, 2008).

This microalga improves the hyperlipidemia, it helps in decreasing the hypertension as well as the level of glucose in the body (Spolaore et al., 2006). It also has been reported that apart from this, the other freshwater microalga, that is Chlorella sps., is having the potential to have various numbers of different health benefits such as a reduction of glycemia and cholesterolemia.

These sorts of species could also be used in order to elevate the production of cytokine in order to stimulate immune response (Barrow and Shahidi, 2008). Even if few microalga strains have been approved for human nutrition, few experiments with other microalgae have been directed on the animal models.

Certainly, a lyophylisate of *Porphyridium cruentum* strain have been utilized in the form of a food supplement for Syrian golden hamsters. With respect to this experiment, it has been revealed that the application or usage of this microalga biomass will help in decreasing the circulating cholesterol in the body (dose-dependently) as well as body fat (expressed in the form of percentage) in hypercholesterolemic animals (Harding et al., 2009).

Figure 4.7: Health benefits of microalgae in preventing cardiovascular diseases.

With respect to the diabetic rats, it has been displayed that groups that are fed with the microalga *Isochrysis galbana* shown a decrement in the level of glucose as well as a decrement in the lipid values (triacylglycerol and cholesterol) along with some weight loss. With respect to the lipoprotein metabolism, this microalga provides assistance in elevating the low-density lipoproteins and helps in reducing the concentrations or content of the high lipoprotein (Nuno et al., 2013).

4.11. MICROALGAL COMPOUNDS WITH ANTI-INFLAMMATORY PROPERTIES

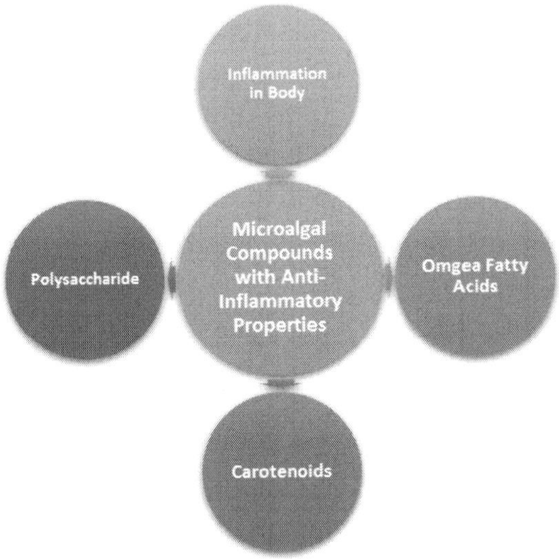

Figure 4.8: Microalgal compounds with anti-inflammatory properties.

4.11.1. Inflammation in the Body

Inflammation is a natural defense mechanism against pathogens. Never the less, the chronic episodes of inflammation may produce renal, neurodegenerative, as well as cardiovascular diseases. In addition to this, the chronic episodes of inflammation also leading to cause of cancer, diabetes, along with the obesity.

There are growing indications and proofs that the oxidative stress is the common factor underlying several numbers of inflammatory pathways, for example, an increment in the reactive oxygen species (ROS) has been recommended to be directly involved in Alzheimer's disease.

This is aided with the help of the overproduction of reactive oxygen species in early-stage patients that outcomes in cleaving and misfolding of the amyloid precursor protein to amyloid beta-peptide (Aβ), which encourages the mitochondrial dysfunction and enables or simplifies the formation of amyloid plaques.

The disease pathology upregulates proinflammatory cytokines as well as free radical generation, which as an outcome stimulates the production of Aβ peptide along with the other inflammatory pathways.

In addition to this, reactive oxygen species which is generated in brain tissues can modulate synaptic as well as non-synaptic communication between neurons that as an outcome, may cause neuroinflammation and cellular death, and in the end, neurodegeneration and loss of memory.

Outcomes from the epidemiological as well as experimental studies strongly recommends that the utilization of the medicinal plants or dietary supplements along with the bioactive plant-based extracts can have anti-inflammatory effects. Contradictory outcomes have also been reported.

In the same way, natural products that are made up with the use of microalgae have been investigated for their anti-inflammatory properties. The most studied compounds are carotenoids, polyunsaturated fatty acids, and modified carbohydrates.

Never the less, still, there are several numbers of other microalgal metabolites that are also stated to suppress chronic inflammation and alleviate disease-associated symptoms, which is consisting of phycobiliproteins, phenolic compounds, and several terpenoids.

The extracts of the algae have shown anti-inflammatory activities, considerably with the help of inhibited production of proinflammatory cytokines and eicosanoids and the lessen expression of proinflammatory genes. The mechanisms of the action from these algal metabolites are very different and are consisting of the modulation of enzyme activities (e.g., phospholipase A2, cyclooxygenase-2 (COX-2), nitric oxide synthase (NOS)), regulation of cellular activities, and the interference of two major signaling pathways.

These are consisting of the nuclear factor κB (NF-κB) and the mitogen-activated protein kinase pathways, which plays an important role in the production of the several numbers of different proinflammatory mediators.

The algal compounds that are present in the extracts may also have the anti-inflammatory characteristics that are integrated with their antioxidant properties.

4.11.2. Omega Fatty Acids

Long-chain polyunsaturated fatty acids, which are consisting of EPA and DHA, are shown to have a therapeutic role in a variety of inflammatory pathologies, for example, Alzheimer's disease, arthritis, and lupus. An increasing number of studies in animal and in vitro trials support the anti-inflammatory properties of microalgal oils.

According to the reports of the well-known researcher named as Nauroth, stated that the application of docosapentaenoic acid (DPA), another omega-3 polyunsaturated fatty acids, from Schizochytrium sp. inhibited the lipopolysaccharide (LPS)-stimulated secretion of interleukin (IL)-1 beta and tumor necrosis factor-alpha (TNF-α) in the human peripheral blood mononuclear cells.

The authors also found that rats fed an algal oil which are consisting of docosapentaenoic acid (which is 16% of the total fatty acids) and DHA (which is 40% of the total fatty acids) considerably decreased inflammation in paw edema model rats in comparison to the control.

According to a well-known researcher named Banskota, also reported that the anti-inflammatory impacts in RAW 264.7 macrophage cells from lipid extracts which are consisting of monogalactosyl diacylglycerol from *Tetraselmis chuii*, *Chlorella sorokiniana*, and *Chondrus crispus*.

In addition to this, the application of the microalgal oils in the diet, more particularly, those that are rich with omega-3 fatty acids, has been showed to bring the chemopreventative impacts on the azoxymethane-induced colonic aberrant crypt foci functions.

Macroalgal lipids have also been discovered for their effects, and as per the reports of researcher, named Khan stated that the anti-inflammatory characteristics of omega-3 fatty acids from the edible brown seaweed, Undaria pinnatifid; stearidonic acid (SDA) was active against the phorbol myristate acetate-induced ear inflammation. On the other hand, EPA was active against edema, erythema, and blood flow in mice.

According to the reports of the researcher named Vo, stated that the polyunsaturated fatty acids from the brown seaweed, Ishige okamurae, were played in an important role in increasing the inflammation because of an allergic response by decreasing histamine release and modulating inflammatory cytokine production in human basophilic KU812F cells.

In this way, it is recommended that the extracts of the lipids that are rich in polyunsaturated fatty acids (from microalgae or macroalgae) could be very advantageous as functional food ingredients in control inflammation.

4.11.3. Carotenoids

Carotenoids are well known for their anti-inflammatory characteristics. On the other hand, they are structurally related, they can have various mechanisms of action. More particularly, astaxanthin has received a significant amount of attention because of its anti-inflammatory characteristics.

The outcomes from *in vitro* studies and *in vivo* studies represents that the astaxanthin, which is produced with the help of the microalga, *Haematococcus pluvialis*, possessed inhibitory impacts on LPS-induced inflammation.

In the matter of fact, the anti-inflammatory impacts of astaxanthin (dosage of 100 mg/kg) were more pronounced than that of a common anti-inflammatory drug, prednisolone (dosage at 10 mg/kg).

The inhibitory activities of xanthophyll carotenoids have been showed for inducible cyclooxygenase-2 (COX-2) or nitric oxide synthase (NOS) enzymes in cell lines (e.g., BV2 microglial cells and RAW 264.7 macrophage cells) that were stimulated with the help of LPS.

Astaxanthin has been stated, in order to inhibit the production of nitrous oxide (NO), prostaglandin E2 (PGL2), and in order to lower the levels of proinflammatory cytokines, TNF-α, IL-1β, and IL-6. Moreover, treatment with the help of astaxanthin at 5 mM was found to enhance the phagocytic as well as microbicide function, and in order to reduce the oxidative harms to the lipids and proteins in human neutrophils.

In addition to this, cells pre-incubated with astaxanthin restored expression levels of SHP-1, a small heterodimer protein tyrosine phosphatase, and decreased the expression of NF-κB.

Outcomes that are collected from the animal studies were in respect with those of in vitro trials. According to the reports of the well-known researcher, named Bennedsen stated that the astaxanthin from *H. pluvialis* at 200 mg/kg decreases the levels of gastric inflammation in Helicobacter pylori-infected mice following a treatment interval of 10 days.

The decrement in the inflammation and the decrease of mucosa-related bacterial loads in the stomach of infected mice were attributed to the action of astaxanthin in order to block the secretion of the interferons, which boosts IL-4 that are release in splenocytes.

4.11.4. Polysaccharides

Algae synthesize several numbers of various carbohydrates which are having different structure with various kinds of glycosidic bonds. They can be categorized in the form of homo-polysaccharides and hetero-polysaccharides, sulfated polysaccharides (SPS), glycoproteins/peptides, and pectin.

Never the less, the relationship between the structures of polysaccharides and their biological functions are not well categorized, and the reason behind this is the variety and complicatedness of these sorts of polymers, namely the absence of monomeric and structural information.

Sulfated polysaccharides (SPS) from microalgae and macroalgae are the most studied bioactive group of carbohydrates with are having anti-inflammatory activity.

These sorts of compounds can fluctuate extensively in their composition, that is, kinds of monosaccharides, and their level of sulfation. Various kinds of microalgal genera have been discovered for their anti-inflammatory characteristics, which are consisting of Porphyridium (which is Rhodophyta), Phaeodactylum (which is Heterokonts), along with Chlorella (which is a Chlorella).

As per the reports of the researcher, named, Guzman, stated that crude polysaccharide which is extracted from the *Chlorella* stigmatophora and *Phaeodactylum tricornutum* showed an anti-inflammatory response in a paw edema test.

4.12. FUTURE PROSPECTS

With respect to the growth of the population, all across the globe, and the availability of the terrestrial food products, microalgae can work as a reliable and sustainable alternative to the commonly used supplies from the animal or plant origins.

Never the less, at present interval of time, only a limited number of strains, primarily the cyanobacterium, Arthrospira, and few green microalgae (e.g., Chlorella, Dunaliella, and Haematococcus), have been used in the commercial production of a limited number of products.

The change from a niche market to the extensive application of products that are made of algae in the form of food supplies will necessitates a broad research and development. This would contain the screening of new species or the enhancement of the existing strains with the help of genetic engineering and alterations towards the development of the microalgae with improved production of the targeted metabolites.

Eventually, it is very important that microalgal cultivation and processing can be done economically in order to make sure a cost-effective production that can compete with synthetic or biological production from the other organisms.

There has been an increment in the demand for functional foods and nutraceuticals with targeted health benefits. With respect to this, the utilization of protein hydrolysates and bioactive peptides has acquired interest for the prevention of illness and the improvement of the symptoms from various kinds of diseases.

Although, there are not so many studies on peptides of microalgal origin are available, but some of them which are available have provided the proof of promising *in vitro* biological characteristics.

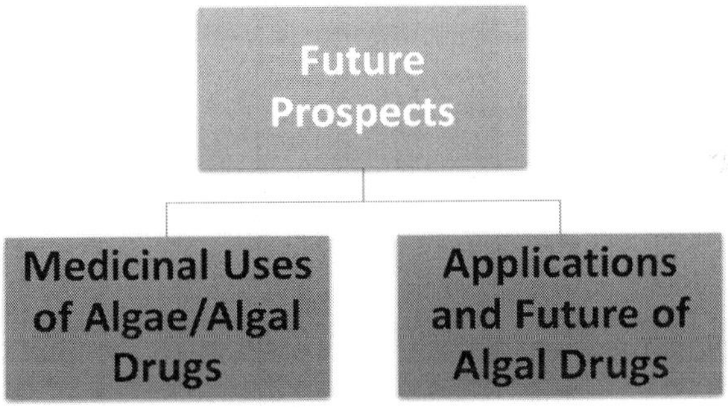

Figure 4.9: Future prospects.

Never the less, more in vivo and clinical trials are required in order to find out the bioavailability of peptides after oral administration, more particularly since loss of bioactivity may take place because of the hydrolysis and/or absorption of peptides in the gastrointestinal tract.

In order to maintain the bioactivity of peptides, various sorts of different strategies are being used, for example, microencapsulation, nanoencapsulation, or chemical modification.

4.12.1. Medicinal Uses of Algae/Algal Drugs

Several numbers of kinds of algae have various medicinal properties, which makes them distinct from the other. They are used for different treatments and below is a small list of these sorts of algae along with their medicinal cures;

1. Enteromorpha: It can be utilized in order to treat hemorrhoids, parasitic disease, goiter, coughing and bronchitis; fever reduction capacity and ease pain.
2. Acetabularia: This can be utilized in order to treat urinary diseases and edema.

3. Laminaria: It can be utilized in order to get rid of thyroid problems along with the urinary diseases.
4. Sargassum: It can be utilized in order to treat cervical lymphadenitis, edema; diminishes inflammation; induces urination; contains both iodine and potassium.
5. Gelidium: It can be utilized in order to extract agar.
6. Corallina: It can be utilized in order as pesticides.
7. Grateloupia: Blood sugar lowing capability.
8. Gloeopeltis: Treatment for tonsils, goitre.

4.12.2. Applications and Future of Algal Drugs

Algal drugs are having several numbers of different applications, which make it a new boon for the future in drug and pharmaceutical sectors. The applications are consisting of;

1. High value oils;
2. Cosmetics;
3. Colorants;
4. Wastewater treatment;
5. Food supplements;
6. Personalized drugs;
7. Fertilizers;
8. In forensic medicines.

In this way, as scientists are looking for a biological drug that are less expensive or cost-effective, green algae have been the upcoming trend, and looking at the above applications, it is becoming quite friendlier. Seeing the potential of bioactive compound production, more particularly with the help of green algae, is a boon to the pharmaceutical research.

4.13. CONCLUSION

In the conclusion of this, the marine microalgae in the food and health industry are generally used in the current scenario to deal with the problems and the challenges being faced regarding these industries. The microalgae play a vital role in the food as nutraceuticals as it contains a sufficient amount of nutrition which are required by the body.

The demand of microalgae in the food industry these days are rapidly risingrising due to the various advantages. The concept of sustainable production is somewhere fulfilled with the help of the microalgae. Therefore, in the current scenario, the microalgae are proving to be a boon. Microalgae and cyanobacteria are also playing important role within the health industry.

Due to the various reasons serious diseases such as heart and hypertensive could be easily caused in order to deal with such harmful diseases the microalgae are used and it has also shown tremendous results. There are many properties being possessed by the microalgae it acts as anti-inflammatory components which helps to reduce the inflammation in the body. Also, there is a high level of omega 3 present in the microalgae which is very good for the healthy hair growth. There are various medicinal uses of the microalgae.

REFERENCES

1. Barkia, I., Saari, N., & Manning, S., (n.d.). *Microalgae for High-Value Products Towards Human Health and Nutrition.* [online] Available at: https://www.ncbi.nlm.nih.gov/pmc/articles/PMC6562505/ (accessed on 31 July 2021).
2. Bassconnections.duke.edu. (n.d.). Marine Microalgae For Sustainable Production of Food And Fuel (2018–2019) | *Duke Bass Connections.* [online] Available at: https://bassconnections.duke.edu/project-teams/marine-microalgae-sustainable-production-food-and-fuel-2018–2019 (accessed on 31 July 2021).
3. *Microalgae and Its Use in Nutraceuticals and Food Supplements.* (n.d.). [online] Available at: https://www.researchgate.net/publication/337700731_Microalgae_and_Its_Use_in_Nutraceuticals_and_Food_Supplements (accessed on 31 July 2021).
4. Chacón-Lee, T., & González-Mariño, G., (2010). Microalgae for "Healthy" Foods-Possibilities and Challenges. *Comprehensive Reviews in Food Science and Food Safety,* [online] *9*(6), 655–675. Available at: https://onlinelibrary.wiley.com/doi/full/10.1111/j.1541–4337.2010.00132.x (accessed on 31 July 2021).
5. Matos, Â., (2017). The Impact of Microalgae in Food Science and Technology. *Journal of the American Oil Chemists' Society,* [online] *94*(11), pp. 1333–1350. Available at: https://www.researchgate.net/publication/320046902_The_Impact_of_Microalgae_in_Food_Science_and_Technology (accessed on 31 July 2021).
6. Mimouni, V., Ulmann, L., Haimeur, A., Guéno, F., Meskini, N., & Tremblin, G., (2015). Marine microalgae used as food supplements and their implication in preventing cardiovascular diseases. *OCL*, [online] *22*(4), p.D409. Available at: https://www.ocl-journal.org/articles/ocl/full_html/2015/04/ocl150006/ocl150006.html (accessed on 31 July 2021).
7. Mourelle, M., Gómez, C., & Legido, J., (2017). The Potential Use of Marine Microalgae and Cyanobacteria in Cosmetics and Thalassotherapy. *Cosmetics*, [online] *4*(4), p.46. Available at: https://www.researchgate.net/publication/320793037_The_Potential_Use_of_Marine_Microalgae_and_Cyanobacteria_in_Cosmetics_and_Thalassotherapy (accessed on 31 July 2021).

8. Nilesh Hemantkumar, J., & Ilza Rahimbhai, M., (2020). Microalgae and Its Use in Nutraceuticals and Food Supplements. *Microalgae – From Physiology to Application*, [online] Available at: https://www.intechopen.com/books/microalgae-from-physiology-to-application/microalgae-and-its-use-in-nutraceuticals-and-food-supplements (accessed on 31 July 2021).
9. Pooja, S. (2014). Algae used as Medicine and Food-A Short Review. 1st ed. [ebook] Tamil Nadu, India: *Journal of Pharmaceutical Sciences and Research*, 33–35. Available at: https://www.jpsr.pharmainfo.in/Documents/Volumes/vol6issue01/jpsr06011408.pdf (accessed on 31 July 2021).
10. Aditya, T., Bitu, G., & Mercy, G. (2016). *The Role of Algae in Pharmaceutical Developmen*t. [online] Rroij.com. Available at: https://www.rroij.com/open-access/the-role-of-algae-in-pharmaceutical-development-.php?aid=79686 (accessed on 31 July 2021).

Chapter

5

Ecology and Management of Aquatic Plants

CONTENTS

This chapter revolves around the ecology and management of aquatic plants, which starts with the varieties of plants that is free-floating, emergent, and submergent. Then it explains the ecology of aquatic plants, classification of macrophytes, the challenges and adaptions to their life present in water. It also sheds light on the different types of biotic interactions and aquatic management, the plans and strategies involved in the aquatic management.

It also emphasizes on the long-term management, which goes on with proper planning and treatments along with explaining the short-term management of aquatic plants, the control methods for aquatic plants in ponds, then goes on defining the various aquatic control methods and the general principles for the cultivation of aquatic plants.

5.1. INTRODUCTION

Our aquatic permaculture systems take into consideration making an assortment of specialties to completely adjust, upgrade and acquire plenitude through edge plan in three measurements and system mix. The specialty creation should be possible in practically all life realms, yet the primary portion of the part will clarify the fantastic measure of biodiversity that is held inside the four significant aquatic plants zone.

Those zones all have their capacities and interrelate with other life structures like microorganisms, warm-blooded creatures, reptiles, and fish. Through strong plan the accompanying four aquatic plant zones can be fused:

1. Free-floating plants;
2. Emergent plants;
3. Submergent plants; and
4. Floating leaved plants.

The accompanying outline shows these zones and an ideal circumstance for every one of the specialties and their spatial connections inside the aquatic system. These zones can thrive and relying upon your aquaculture wants, your plan will direct the state of the lake and how much space you devote to any of these specialties. Recall that aquatic systems are similar to food woods, there is a progression that happens so the system isn't static, rather very powerful.

5.2. FREE FLOATING PLANTS

Perhaps the most noticeable and energizing layers to communicate with is the one that buoys unreservedly on the water's surface, roots plunging underneath yet discovering no anchor in soil. They move around with flows and wind and have shifting sizes, development rates, and employment.

In like manner they do cover rapidly, subsequently delivering a plenty of biomass. Frequently marked as invasives, they do quickly and productively cycle overabundance supplements which appear to be in plenitude due to poor farming practices and metropolitan arranging that drain nitrogen and phosphorous overflow.

These can be cycled to construct soil and in drier areas discovering green material for summer treating the soil and compost can be scant however, aquaculture frameworks can sparkle their abundance and still, after all that.

The principal such plant to analyze is the universes littlest blooming plant-duckweed or Lemna. It covers the water surface rapidly and totally with its little green leaves. In sentiment dialects they are classified "water lentils" with this interpretation showing their protein potential.

Consequently, the duck part of its name as ducks truly appreciates benefiting from this plant that harbors 43% unrefined protein when dried. It tends to be scooped out and taken care of two chickens also, and with its rich lysine content, it very well may be the distinction creator in a chickens diet for delivering brilliant eggs rather than light yellow yolks.

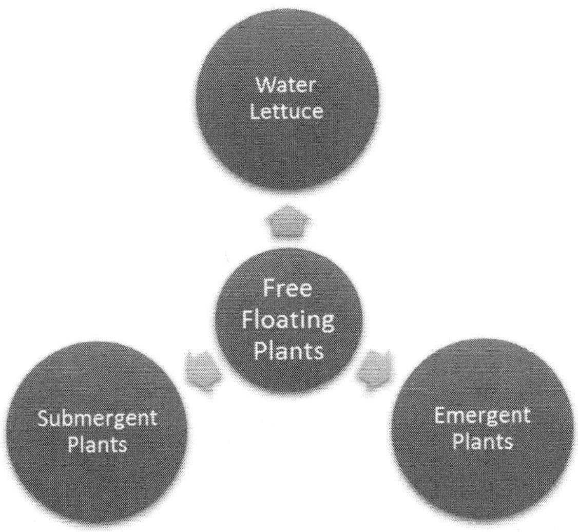

Figure 5.1: Types of free-floating plants.

It likewise can be scooped off and be a mulch plant or included manure heaps. I have additionally seen it be utilized to cover water tanks in Portugal to stifle dissipation and mosquito multiplication. These plants are local everywhere on the world, and their capacity to replicate their mats of surface vegetation in one day in the wake of scooping half off is an incredible indication of bounty.

5.2.1. Water Lettuce

Water lettuce, while not consumable for us, is very valuable in its tropical (or developed as a yearly) growth regions for in general system wellbeing. Ducks love to benefit from its new growth while numerous insects and mollusks discover space to flourish among its leaves and broad root systems.

It holds altogether more biomass than duckweed making it a shockingly better decision for biomass creation. It roots likewise house sifting microorganisms. To get this one set up, you may need to put a fence around a corner patch as the ducks will pound it back. Joined with fish snacking its roots, it tends to be difficult to build-up due to its lettuce like quality to other people.

Another tropical decision of free-floating aquatic plants is the water hyacinth. With its waxy leaves, verdant tall, algae, and wonderful lilac blossoms, it's a plant of incredible magnificence. Anyway, it is highly censured in light of the fact that it is very savage in its development, totally stifling out streams rapidly.

Figure 5.2: An illustration of water lettuce.

Source: Image by Wikimedia Commons.

This is obviously a reaction to nutrification from agrarian, mechanical, and district spillover. It's an incredible filter plant and offers an environment to numerous bugs and mollusks in this way assisting with setting up higher types of life.

In Costa Rica, an abundance of harvest of a bumper cacao crop from others mulching with water lettuce and water hyacinth the prior year. These supplement and mineral rich plants gave the plants precisely what they expected to thrive.

Moving the plant material to the cacao trees was simple as they were in relative area. Accordingly, choosing what trees encompass your oceanic lakes is a significant piece of hydroponics plan joining into a general site plan.

Another dazzling free floater that develops more like duckweed is the tangling plant of *Azolla azolla* or Water Fern. Its leaves take on a cypress/ cedar look in structure; however, eventually, in the year, they frequently become red, giving the water body a striking shading distinction between the adjoining green growth of the land.

This plant grows rapidly and is high in nitrogen because of its microbial affiliations. It fills comparable natural specialties and soil building attributes as the others above yet unquestionably recognized by its conspicuous antiquated structure, stylish upgrade, and nitrogen-fixing.

5.2.2. Emergent Plants

The next gathering of plants being talked about is those that develop directly at the edge of the lake and into the water body itself.

They are found on the edge where water levels will in general vacillate and can be totally immersed or standing some separation from the water's edge at various seasons.

Basically, they are rooted in the water body yet then arise through the water surface. They are incredible colonizers and help to fuel lake progression with their plentiful natural mater.

They are incredible filter plants and are regularly alluded to as wetlands plants as these shallow kidneys (wetlands) of nature highlight these rising plants broadly.

Figure 5.3: An illustration of emergent plants.

Source: Image by Wikimedia Commons.

Extraordinarily, they do siphon oxygen into their root zones and help to keep vigorous/anaerobic natural matter preparing in balance. Their underlying foundations are accordingly colonized by a wide scope of organisms assisting them with retaining supplements and weighty metals.

The primary plant to analyze and the most widely recognized overall is the Cattail or Typha. It puts on unbelievable growth rates and its starchy tuber spreads rapidly. Its rhizome is cattail squite tough and powers itself more profound into the water and onto the drier banks.

It is edible, and pound for pound (kg for kg), it far outcompetes a potato for each section of land (or hectare) in its rhizosphere yield. Likewise, it very well may be eaten like asparagus in the late winter when its new shoots arise.

Besides its pollen, as found in the beneath picture, is a nutritious enhancement that is added to flour for heating. At that point, to finish that off, its fiber utilizes are huge also. First, the seed heads, which structure after this mineral rich pollen is pollinated, are a phenomenally fine fiber. This gives it unbelievable insulative characteristics and is a top pick of dig for adding to a completion mortar.

The fine fiber helps in this layer for it to not break and substitutes other difficult to come by fine strands from creatures. Likewise, the cattail plant is utilized in weaving customarily for crates, mats, bedding and even the basis of the aquatic mats that made the very useful aquaculture system of Central

Mexico known as Chinampas. Cover material is likewise gotten from this plant which is normal of this plant grouping.

Another new plant and valuable fiber plant are a lot of rushes that can be seen all over the planet. They are regularly utilized for thatching and pfaf. org even says that one Rush Plant species is eatable. They are high in silica, making their cells difficult to separate, which is the thing that makes them an incredible thatch plant yet also in addition a garden mulch when material is scant.

This is one plant that has been decreased in its common territory because of wetlands depleting. However, making edge in lakes and stream circumstances permit this plant to reappear.

Sedges are likewise incredible fiber plants and the exemplary story of making paper from our adolescence on the Euphrates comes from this plant family-the papyrus. There is a fun saying to recognize rushes and sedges, and it goes this way, "sedges have edges, and rushes are round."

So, in the event that you end up attempting to recognize plants on an aquatic edge, a three-sided feel in your grasp demonstrates that could actually be a sedge while rushes will turn with their round shape. Grasses are empty inside, giving you one more ID tip.

Likewise, the pleasant eating water chestnut belongs to this grouping and where suitable ought to be considered for an extraordinary vegetarian yield from aquaculture systems. This tuber is regularly found in oriental cooking and its light flavor and fresh surface is a significant great collect.

5.2.3. Submergent Plants

The following gathering of plants to be analyzed is those plants that are rooted in the lake base and send their leaves upwards however don't break the water surface other than their blossoms on certain species. They structure thick ocean growth like mats as compare to "backwoods of lakes."They harbor a wide range of bugs and mollusks and are an extraordinary shelter for little fish. This obviously prompts predation of different kinds whether from crawfish-, fish-, or fish-eating birds. Since they don't break the water surface, the plants can have their stoma more open in this way delivering piles of oxygen into the water body. They have staggering measures of edge from their fluffy leaves and are a serious delightful sight in streams yet frequently viewed as obtrusive. They grow not very deep waters and frequently can be seen from the banks as you glance through the water body.

Figure 5.4: An illustration of submergent plants.

Source: Image by Wikimedia Commons.

Milfoil forms huge colonies of its thick and jungly vegetation under the water surface. Its Coontail creates numerous spaces for animals to flourish as certain animals benefit from its leaves while others feed off the film of microorganisms that exist there.

With hunter fish covering up and prepared to jump, youthful fish dart all through the influencing vegetation. It moves nutrients to plant development, which some call obtrusive, yet is misinformed as we should take a gander at the underlying root cause: manmade spill over contamination.

5.3. MACROPHYTES: ECOLOGY OF AQUATIC PLANTS

Aquatic plants add to keeping up key capacities and related biodiversity in freshwater biological systems and to give the requirements of human social orders. The way the natural specialties of macrophytes are dictated by abiotic channels and biotic ones is thought of.

A basic, extensively appropriate model of the conveyance of growth structures as indicated by abiotic channels is proposed. The ramification for the elements of plant communities and the principal dangers to macrophyte occurrence and variety are examined.

Macrophytes are essential makers and give environments and shelters to periphyton, zooplankton, other invertebrate species, fish and frogs. They likewise play key capacities in biogeochemical cycles (e.g., natural carbon creation and phosphorous assembly) straightforwardly affecting hydrology and dregs dynamic of freshwater biological systems through their consequences for water stream. Macrophytes add to keep up key capacities and related biodiversity in freshwater environments, yet additionally through the administrations they give straightforwardly to man. Just submerged macrophytes that live forever or more often than not submerged in water are thought of, as such species have rather differentiating environments and explicit necessities contrasted with helophytes (e.g., part of the way arose macrophytes). As far as territories, the attention will be on regular environments of freshwater macrophytes, that is, mainland standing and running biological systems, as lakes, lakes, wetlands, waterways, and streams. Types of salty waters are not thought of, as the necessities and stress that influence seagrasses part of the way vary from those influencing freshwater species.

5.3.1. Classification of Macrophytes

Macrophytes incorporate bryophytes (greeneries, liverworts and hornworts), pteridophytes (plants and greenery partners) and spermatophytes (seed-bearing plants) whose photosynthetically dynamic parts are forever submerged in water or which skim on the outside of water. Under 2% of vascular plants are aquatic. There are comparable extents of aquatics in monocots and dicots at the level of the superorder, however relatively a larger number of aquatics in the monocots than in the dicots at the family level (35% of monocot families present amphibian species, yet just 12.1% of dicot families).

Figure 5.5: An illustration of bryophytes.

Source: Image by Wikimedia Commons.

A couple of groupings of Angiosperms are completely aquatic, with respect to model, Nympheales, Hydrocharitales, Zosterales, Alismatales and Podostemales. Most other aquatic species have a place with bunches where just a few families or a few genera are aquatic. The topic of the aquatic character of the main angiosperms has been intensely discussed, yet so far has not been clarified.

The new angiosperm characterization diagrams the basal ancestry of a few aquatic taxa (e.g., Ceratophyllales, Nymphaeaceae and Alismatales; Angiosperm Phylogeny Group, 2003), and fossil records propose the early colonization of oceanic environments by angiosperms. Furthermore, angiosperms had the option to return a few times to water over the span of advancement.

5.3.2. Challenges and Adaptations to Life in Water

An angiosperm needs to tackle a few issues to effectively colonize aquatic living spaces. The low-dispersion pace of gases (and all the more especially carbon dioxide and oxygen) prompts issues of carbon and oxygen supply.

The assembly and transport of nutrients in the plant are further limitations, as happening from leaves can presently don't drive interior water transport in aquatic conditions, with the exception of plants with drifting leaves. Among factors that can act emphatically on plant development, the high thickness of water (in excess of multiple times higher) contrasted with air permits plants to be precisely upheld.

Supplements happen in water, in touch with the over-the-ground organs of the plants and are conceivably accessible for plant development. aquatic plants present a typical arrangement of characteristics that adjust them to such natural surroundings.

Among them, the event of aerenchyma expands oxygen motion from shoots to roots. A more prominent leaf surface (when similar species can create both arose and lowered leaves, the leaf surface is unequivocally more noteworthy for lowered leaves contrasted with arose ones), along with a thin cuticle, increment contacts with water and carbon take-up.

Water plants are normally ineffectively lignified, as water shields plants from gravitational pressure, and xylem cells are supplanted by a couple of xylem lacunae, with a lot higher pressure driven conductance.

The event of a Casparian (band of cell divider material made out of lignin or potentially suberin, in the spiral and cross over dividers of the endodermis) in the endodermis of the tail and the leaves is in this setting saw as a pressure driven cover, saving the detached dispersion of supplement arrangements towards the aerenchyma, and proposing that root pressure is the significant main thrust for water and nutrients from roots to apical meristems.

Helpless oxygen accessibility submerged represented the development of huge aerenchyma all through the plant to improve the photosynthetic oxygen transition from shoots to establishes covered in anoxic residue and to the adjoining rhizosphere.

5.4. BIOTIC INTERACTIONS

5.4.1. Competition for Light

Rivalry measures happen both inside macrophyte networks and among macrophytes and microphytes. In earthbound territories, the primary boundary that decides serious capacity for light is plant size, which is emphatically associated with plant biomass.

In aquatic living spaces, the capacity to reach light isn't straightforwardly identified with plant size, yet to their capacity to create produce floating or emerged leaves. Moreover, the size of plants isn't just identified with their biomass, since lowered organs (e.g., leaves) present lower development costs than floating ones.

As a result, species that present exceptionally differentiating development structures are considered as seriously prevailing: free-floating species and tall species with floating leaves. Among different groupings, species ready to create through plastic changes, either emerged or floating leaves (e.g., numerous Alismataceae), are more ready to endure contest pressure.

Evergreen and wintergreen species (rosette and little caulescent species) are likewise ready to get away from contest from summer greens (e.g., tall species with floating leaves and free-floating species) by developing during winter in environments where there are diminished occasional varieties of temperature.

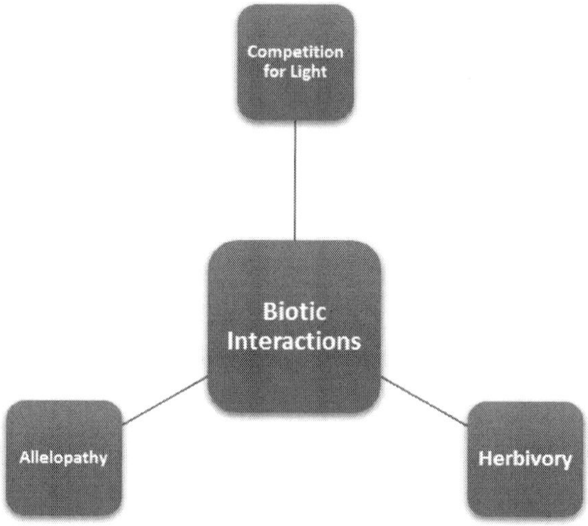

Figure 5.6: Biotic interactions.

5.4.2. Allelopathy

Allelopathy, that is, the creation and arrival of chemical substances that hinder the development of other macrophytes or microphytes, has been exhibited to happen among macrophytes, despite the fact that many distributed investigations neglect to show in an unfalsifiable manner the allelopathic system, and the poisonous compound associated with the connection.

These creators give a survey of the applicable papers that recognize allelopathic species and show that allelopathic impacts of macrophytes happen in differentiating scientific classifications like Najadaceae or Potamogetonaceae. In certain circumstances, such systems block microphyte development and keep an unmistakable water stage in the lake.

5.4.3. Herbivory

Aquatic plants are devoured by numerous creatures, like fishes, birds, warm blooded animals, molluscs and arthropods (creepy crawlies, crustacean), and plant populaces can likewise be altogether influenced by brushing. Some slow eaters practice on one macrophyte animal varieties (e.g., the coleoptereae *Galerucella nymphaeae* that eat Nuphar leaves), however a large number of them are polyphaeous.

The reasonableness of aquatic plants to brush relies upon scientific categorizations (Elger et al., 2004), yet in addition on the formative stage, the lowered versus drifting sort of leaves (Sand-Jensen and Jacobsen, 2002), and on the organ considered, a few animal groups touching explicitly blossoms, and seedlings being more defenseless to brush than completely created plants.

The event of a cuticle, or reinforcing tissues, diminishes the affectability of aquatic plants to touch (Elger and Willby, 2003). Aquatic plants likewise produce numerous chemical mixtures (tannins, phenols, alkaloids), and it has been recommended that some of them could be allelopathic.

5.4.4. Other Biotic Interactions

Aquatic plants can be colonized by mycorrhizal parasites (Sraj-Krzic et al., 2006). The colonization levels of certain species ready to become apathetically submerged or as a riparian animal types have been exhibited to be higher in earthly circumstances, and profoundly associated with plant phosphorous fixations, recommending a positive job of such organisms for plant phosphorous take-up that could be expanded in dry conditions, despite the fact that this constructive outcome isn't critical for all species contemplated.

5.5. AQUATIC PLANT MANAGEMENT

Aquatic plant management includes the science and approaches used to control obtrusive and non-intrusive aquatic plant species in streams. Techniques utilized incorporate spraying herbicide, natural controls, mechanical evacuation just as territory alteration. Forestalling the presentation of obtrusive species is ideal.

Aquaculture has been a source of outlandish and eventually obtrusive species presentations such *Oreochromis niloticus*. Aquatic plants delivered from home fish tanks have additionally been an issue.

5.6. THE AQUATIC PLANT MANAGEMENT PROGRAM

The objective of any plant management program ought to be to keep an appropriate equilibrium of plants inside a lake. Preferably, every aquatic plant management program will have two stages: (1) long-term management (supplement control); and (2) short-term management (direct control of macrophyte and green growth populaces). Short-term management is moderately simple to carry out, however long-term management is more muddled. It requires significant community indulgence and collaboration, and results require years as opposed to days to create. The rest of this notice will think about the standards and strategies of long term and short-term aquatic plant management. Few out of every odd rule or strategy introduced will be appropriate to each lake, and surprisingly, those that are ought to be assessed cautiously prior to utilizing them. While financial aspects must, obviously, be thought of, environmental qualities ought to get prime thought before any administration procedure is utilized. Thoughtfulness regarding biological qualities may bring about a program that is less expensive over the time of 10 years or two.

5.7. LONG-TERM MANAGEMENT (NUTRIENT CONTROL)

Aquatic plants require nutrients for growth and propagation. The nutrients regularly viewed as in short supply, and thusly restricting plant growth, are phosphorus and nitrogen. Since aquatic plant growth is straightforwardly subject to the measure of nutrients accessible, nuisance growth is an overall manifestation of high nutrient levels. This is significant, as very regularly, aquatic plant control programs are coordinated uniquely at the aquatic plants, and not at what causes the unnecessary growth (nutrients). A successful aquatic plant management program should give appropriate thought to the measure of nutrients entering the lake. Aquatic plant management strategies planned uniquely to “kill weeds” should be viewed as transitory corrective measures to decrease the side effects of high nutrient levels. Besides, this method proceeds with the cycle of nutrient addition, from the rotting plants, for future, unnecessary, plant growth.

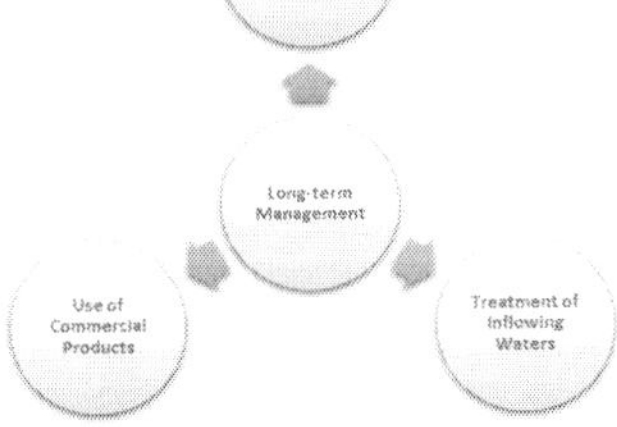

Figure 5.7: Long-term management (nutrient control).

It is more reasonable to control the growth of nutrients from the watershed than to endeavor remedial action after nutrients have entered the lake. Restricting the growth of nutrients into water bodies requires management of nutrient sources.

Regular sources of nutrients are those which would enter a lake, typically in limited quantities, without human impact. Notwithstanding, social sources of nutrients are normally enormous in volumes just as fixations, and significantly speed up the pace of eutrophication. Some nutrient sources, both normal and social.

All nutrient sources will have various degrees of reasonability. Some might be uncontrollable, while others might be controlled with little exertion or cost. In a perfect world, it is attractive to realize which sources are contributing nutrients to a lake and in what amounts. It is then conceivable to change assets and exercises to control nutrient sources to most successfully diminish the measure of nutrients entering the lake.

The best and ideal opportunity to start a nutrient control program is before aquatic plants have accomplished nuisance levels. The administration of nutrient sources is an ongoing duty, which should be heightened as advancement of the watershed proceeds. Strategies for nutrient source management include:

1. appropriate land use planning and zoning,
2. shrewd shopper utilization of commercial items,
3. treatment of inflowing waters high in nutrients,
4. redirection of water high in nutrients, and
5. metropolitan and industrial wastewater treatment.

5.7.1. Proper Land Use Planning and Zoning

Planning and Zoning (P&Z) are public strategies and laws used to manage the utilization of land by local units of government. There are numerous objectives to P&Z, yet fundamentally it is utilized to keep away from land use clashes, guarantee similarity with local area attributes, and ensure general wellbeing, security, and government assistance.

Throughout the long-term P&Z has developed to incorporate different destinations like ecological security. To meet the goals through P&Z, neighborhood statutes are set up to incorporate explicit land-use practices and ought to incorporate consistency with state guidelines where pertinent.

5.7.2. Wise Consumer Use of Commercial Products

At this point, it is notable that cleansers and fertilizers can contribute critical measures of nutrients to our waterbodies. Be that as it may, appropriate use and management of these items, just as elective decisions in cleansers, would considerably diminish the stacking of these nutrient sources. In particular, waterfront landowners should take exceptional consideration in the utilization of cleansers and fertilizers.

To diminish phosphorus stacking into streams, high phosphate cleansers have been restricted in Michigan and a few encompassing states. Moreover, a phosphorous limitation law was passed in Michigan with respect to fertilizers.

Numerous waterfront landowners favor a decent green lawn trimmed right to the edge of the water. Notwithstanding, a maintained yard is likewise one of the essential drivers of exorbitant nutrients to Michigan's water bodies because of fertilizers overflow and the disintegration of properties at the water's edge. A cut grass doesn't channel spillover or hold soil set up the manner in which a very much vegetated buffer of local plants will.

On the off chance that yards should be treated, soils ought to be tried to figure out which substance nutrients are required. In the event that the soil doesn't need phosphorus, a compost with practically no phosphorus ought to be utilized. County cooperative extension agents can give data on soil testing systems and the best strategies for applying composts.

Obviously, the most normal compost, just as least demanding and least expensive, is basically leaving the grass clippings set up. This gives the fundamental nitrogen to new growth through regular breakdown.

Even better, local arranging could be joined into one's yard. Along the water's edge, this contributes not exclusively to nutrient treatment and soil steadiness (further examined beneath), yet is advantageous to local pollinators, local birds, and other untamed life and fish. Local finishing is nearly upkeep free once it is grounded and can be pretty much as straightforward as a 10-foot-wide support to an intricate lovely nursery setting.

5.7.3. Treatment of Inflowing Waters

An inflowing stream, channel, or overland spillover may convey considerable measures of nutrients and different contaminations gathered from sources like rural exercises, metropolitan stormwater seepage, industrial stormwater, and construction stormwater. While it is consistently most straightforward, least expensive, and best to keep contamination from releasing to a water body, in certain circumstances, it is feasible to treat polluted waters.

5.8. SHORT-TERM MANAGEMENT OF AQUATIC PLANTS

Albeit the underlying and proceeding with period of aquatic plant management ought to be the control of nutrient sources, numerous lakes have such genuine plant issues that momentary administration procedures might be expected to keep up the sporting and monetary interests in the lake.

Likewise, in situations where nutrient control is unrealistic, like shallow repositories on major, horticultural or urbanized stream systems, transient administration practices must be led yearly. Indeed, even in such cases, be that as it may, by no means should the total annihilation of aquatic plants be thought of.

This training is ecologically weak and could have entirely unfortunate results. In certain lakes, it very well might be important to adjust sporting exercises to some degree to suit the lake's condition of eutrophication, as opposed to endeavor to change the lake to fulfill sporting needs.

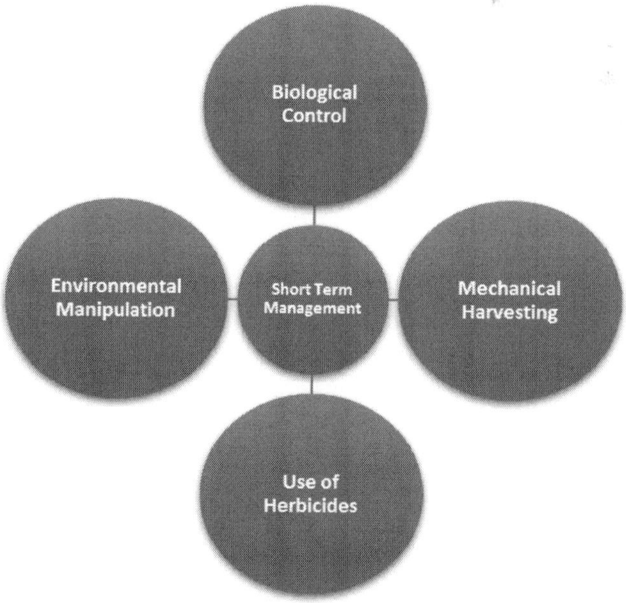

Figure 5.8: Short-term management of aquatic plants.

In circumstances where nutrient control is conceivable, transient administration strategies ought to be viewed as just as transitory measures, intended to supplant irritation plant species with plant species that contention less with sporting and monetary interests.

The momentary strategies for overseeing aquatic plants include:

1. organic control,
2. mechanical gathering,
3. natural control, and
4. utilization of herbicides.

These strategies are coordinated essentially at the outcomes (aquatic plants) of nutrients entering the lake and not at diminishing the progression of nutrients. Sometimes, nonetheless, nutrient levels with the water system might be decreased by chance with specific methods.

5.8.1. Biological Control

Organic control of aquatic vegetation is by and by the most un-comprehended and used of the four short-term the board strategies. Natural control typically incorporates the presentation of a life form that rivals, goes after, hinders the development of, causes illness in, or parasitizes a plant-animal category which has made an issue.

5.8.2. Mechanical Harvesting

Mechanical reaping includes the pulling or cutting and expulsion of macrophytes from chosen spaces of a lake. It might utilize hand devices or exceptionally complex mechanized cutting or rotovating gadgets. The reaping of algae from lakes shows up as of now to be monetarily infeasible basically because of high energy expenses to eliminate the minute plants from water.

Mechanical collecting likewise has disadvantages which should be thought of. It has a high starting venture if an exceptionally produced reaper is bought. A considerable lot of these machines are huge, bulky, and can be harmed by obstacles (logs, stones, and flotsam and jetsam) covered up beneath the lake surface. Also, reaping could help the spread of a plant issue, since parts of specific plants could float into unaffected regions, flourish and develop.

5.8.3. Environmental Manipulation

The target of environmental manipulation is to modify at least one physical or chemical factors basic to plant propagation and development, subsequently making the climate less appropriate to the plant. A few procedures have been utilized with differing levels of progress.

These strategies may not be monetarily or environmentally feasible in each lake. Indeed, even in viable circumstances, a strategy ought to be utilized solely after the specific plant issue, and social and monetary components have been painstakingly thought of environmental manipulation can give some control of aquatic plants, however without a decrease of nutrient inputs, any outcomes accomplished will be just impermanent.

Since a large portion of these strategies are to some degree specialized, just a concise conversation of each is given beneath. The greater part of these exercises requires a grant from EGLE.

5.8.4. Use of Herbicides

Chemical control is another method for briefly controlling aquatic plants and algae. There are various synthetics accessible which offer changing levels of activity time, consistency, cost, selectivity, and security to people, different vertebrates and aquatic creature life.

At the point when herbicides are important for an aquatic plant management program, unique consideration should be taken to secure both the climate and people required, since herbicides are possibly hazardous to both.

5.9. CONTROL METHODS FOR AQUATIC PLANTS IN PONDS AND LAKES

Aquatic plants filling in lakes and lakes are advantageous for fish and natural life. They give food, disintegrated oxygen, and producing and settling natural surroundings for fish and waterfowl. Aquatic plants can trap extreme nutrients and detoxify chemicals. Aquatic wildflowers, for example, the water lily is sold and planted to give botanical excellence to plant lakes.

Nonetheless, thick developments (more than 25% of the surface space) of algae and other water plants can genuinely meddle with lake diversion and undermine aquatic life. Water plants can limit swimming, drifting, fishing, and other water sports.

Water plants can bestow terrible taste (stale smelling flavor), rotting vegetation produces hostile scents (spoiled egg smell), and algae can stain lake waters. Thick developments of plants can cause evening time oxygen exhaustion and fish kills.

Green plants produce oxygen in daylight, yet they devour oxygen around evening time. Disintegrating water weeds can drain the oxygen supply, bringing about sport fish killings from suffocation. Thick plant developments can give an excessive amount of cover, forestalling predation, and prompting hindered (little estimated) sportfish populaces.

5.10. AQUATIC PLANT CONTROL METHODS

Determination of the best treatment or mix of treatments relies upon the types of water plant, the degree of the issue, financial contemplations, nearby natural conditions, and lake uses. In the first place, be certain that you have an aquatic plant issue.

Some aquatic plant developments are minor and brief, and don't need exorbitant weed control activities, in this manner saving you stress, time, and cash. In the event that aquatic plants cover over 25% of the lake surface region, you ought to consider carrying out weed control.

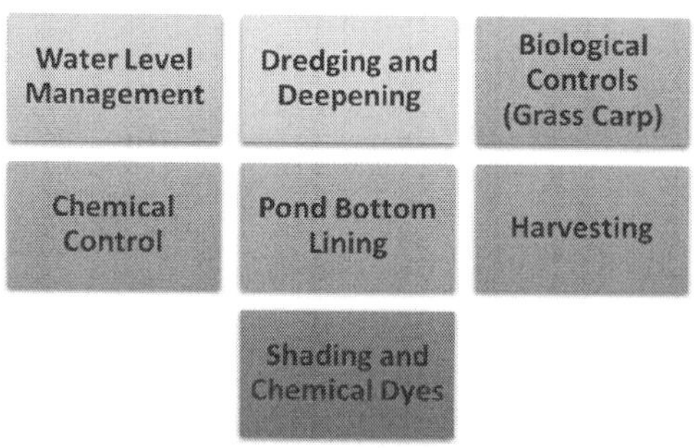

Figure 5.9: Aquatic plant control methods.

Second, various kinds of weeds (green growth, skimming leaf weeds, emanant weeds, and submerged weeds) require various medicines (Shelton and Murphy, 1989). Contingent upon the sort of weed and the seriousness of the issue, one or a mix of the accompanying control strategies can be compelling. The various aquatic plant control methods that are:

1. Dredge and extend the lake;
2. Harvest (manual or mechanical expulsion) weeds;

3. Manipulate water levels;
4. Shade, color;
5. Install lake bottom liners;
6. Use organic controls;
7. Use chemical controls.

5.10.1. Dredging and Deepening

Eliminating lake base residue and building steep lake bank inclines (3:1 slant) are powerful approaches to control established aquatic plants in shallow lakes. Digging decreases aquatic plant issues straight by eliminating the plants, base residue, and related nutrients.

Digging and extending shallow shoreline regions limits weed development by implication by uncovering a dirt layer that might be nutrient-poor or impenetrable to plant roots, and by diminishing the measure of daylight accessible to vegetation.

Digging can be directed on dry sediments, after the lake has been depleted, or submerged. Hydraulic dredging is ordinarily excessively costly for use in little lakes; however, draglines can be utilized for submerged lake digging.

On depleted lake bottoms, dry-land excavation machinery, for example, tractors can be utilized. In spite of the fact that burrowing and eliminating base residue by hand is difficult work, it addresses a basic, prudent, and proficient option to mechanical digging. The dredge spoils ought to be moved beneath the lake bowl to forestall soil overflow once more into the lake.

5.10.2. Water Level Management

Bringing down the water level of a lake can be a simple method to control aggravation oceanic plants. Lake drawdown, especially throughout the cold weather months, opens weeds to cruel conditions, including freezing, desiccation (drying out), solid breeze activity, and bottom sediment compaction.

What's more, ice hurling of the base silt evacuates the weeds and helps in their annihilation. To safeguard effective over-winter control, the base muds should stick to a depth of 4 inches for a little while or more. Overwinter drawdown is particularly successful against cattails, yet some weed species are not constrained by this technique. Vegetation uncovered by bringing down the water level ought to be gathered and taken out from the lake bowl, or the spoiling plants will contribute supplements that advance new developments when the water level is raised.

In lakes without channels, water siphons can be utilized to deal with the water levels. This technique is likewise frequently used to reestablish a hindered fish populace, by driving little bluegill out from the shadows water where they are all the more effectively went after by bass.

5.10.3. Harvesting

Physical expulsion of waterweeds from lakes is a decent control method. Collecting of aquatic plants comprises of three fundamental advances. These are

1. cutting or evacuating the weeds,
2. gathering the cut weeds, and
3. eliminating the weeds from the lake.

Gathering can be cultivated with basic hand devices and actual work or with the assistance of cutting machines. Entire plant expulsion by and large is superior to cutting since certain plants can recreate from cuttings.

In shallow shoreline regions, weeds can be pulled by hand, cut by sickle, uncovered with a tool, and eliminated from the water with rakes or forks. In huge lakes, an assortment of business controlled cutting machines are accessible. Some cutting edges can be mounted on the bow of a speedboat. The achievement of any collecting activity relies upon the brief and complete expulsion of every single cut weed. Aimless or incomplete evacuation of the cut weeds can build the issue, since each unremoved plant part can possibly frame another weed. Additionally, cut plants left in the water will rot and delivery supplements that animate future weed developments. Decaying plants left in the lake use oxygen and can cause fish murders. Gathering gives quick alleviation from nuisance plant developments and doesn't imperil fish life.

5.10.4. Shading and Chemical Dyes

Restricting the measure of daylight accessible to aquatic plants by skimming dark plastic sheeting on the water surface or by utilizing dull shaded colors has been compelling in controlling waterweeds. Dark plastic sheeting appended to Styrofoam floats fills in as a floating conceal which can be moved effectively starting with one spot then onto the next for spot treatment of nuisance water plants in little regions. The floating dark plastic raft ought to stay in one spot for a month to be powerful. This method is ideal for weed control around boat moors, fishing wharfs, and swimming seashores.

Economically accessible nontoxic watercolors (nigrosine, analine, and water conceal) can be utilized to shading the water to diminish light entrance and shade out nuisance plants. To be viable, the colors should continue for half a month.

For the best outcomes, this procedure ought to be utilized in late winter toward the beginning of the developing season before the waterweeds have gotten an opportunity to build up themselves. Applying fertilizers to make algae sprout to conceal lake weeds isn't suggested. This procedure is bound to expand the water weed issue and may bring about harmful sprouts of blue green algae.

5.10.5. Pond Bottom Lining

Covering the base sediments of little lakes with either plastic sheeting, a layer of mineral soils (sand, rock, mud), or both of these materials is a successful waterweed control procedure. Perforated dark plastic sheeting covered with a cover of sand or rock gives a moderately modest option in contrast to digging.

These covers limit the trading of nutrients from the base muds to aquatic plants and restrain the foundation of established waterweeds. Furthermore, dregs cover gives a firm steady base which can limit water loss.

Plastic sheeting (4 mil or thicker) ought to be equitably weighted and punctured with little openings to allow the break of gases delivered by decomposition and to forestall ballooning. In the event that a mineral soil cover is utilized to cover the base dregs, a 6-to 8-inch layer frequently is vital. Covering the lake or lake base residue with a layer of sand, rock, or another mineral soil has demonstrated less powerful than utilizing plastic sheeting in mix with mineral soil. Dark plastic sheeting overlaid with a few creeps of mineral soil is suggested. Significant wetland natural surroundings, for example, fish generating or waterfowl settling regions, ought not be covered.

5.10.6. Biological Controls (Grass Carp)

Introducing animals and plants that eat or rival with waterweeds addresses another control technique. Herbivorous creatures (those that eat plants) incorporate a wide variety of bugs, snails, crawfish, fledglings, turtles, fish, ducks, geese, and swans which can be loaded in lakes to burn-through oceanic plants. Of these, the triploid (sterile fish with half a bigger number of chromosomes than ordinary).

Chinese grass carp (Ctenopharyngodon idella) is a plant-eating fish that can be loaded in lakes to give successful, conservative plant control. The suggested loading rate is around 12 fish (stock enormous fish 9–12 crawls long) per surface section of land.

Lakes with exceptionally thick weeds may require more fish and those with less, less fish. These fish will attempt to move up or downstream out of your lake, so block fish section out of the lake. Most states require a license from the fish and game organization to import and stock these non-local fish.

In Virginia, if it's not too much trouble, call the Virginia Department of Game and Inland Fisheries at (804) 367–1000. One issue with presenting grass carp and different herbivores into lakes for amphibian plant control is that their body wastes go about as fertilizers and facilitate aquatic plant development. Herbivorous creatures, by reusing nutrients, may accomplish more mischief than anything. A wide range of sorts of aquatic plants contend with each other for space, light, supplements, and other basic components. Along these lines, it is once in a while conceivable to debilitate the development of disturbance oceanic plants by planting advantageous ones. Grass carp may eat valuable just as disturbance plants.

5.10.7. Chemical Control

Herbicides (plant poisons) are normally used to oversee land and water plants. Herbicides are moderately simple to apply and might be the solitary viable strategy for control in certain circumstances. Notwithstanding, the treatment of weed-invaded waters with herbicides should be utilized with caution. Herbicides can be harmful to fish and other aquatic life.

Their utilization requires the treated water to be confined from drinking, domesticated animals watering, swimming, fish creation, water system, and different uses until safe levels of the herbicide are reached. Most herbicides are short-lived (10 days or less); however, others are diligent (30-multi day withdrawal from use). Herbicide treatment can be expensive and may give just momentary alleviation from the genuine issue, normally fertile waters. Comprehend that when waterweeds are executed by synthetic compounds, they decay and deliver their contained nutrients (manure) into the lake water. These supplements are then accessible to facilitate future weed development, frequently requiring more treatments.

5.11. GENERAL PRINCIPLES FOR THE CULTIVATION OF AQUATIC PLANTS

To give aquarium plants a totally appropriate medium, we should initially know about the conditions existing in their local living space. Most aquarium plants come from tropical and subtropical regions, with a couple from the hotter pieces of the mild zone.

Regularly the aquarist will find on paper or hear in conversations the suspicion that tropical soils are fruitful and wealthy in nutrients; this is obviously in blunder. Generally speaking, the soil in the jungles is poor since incredible amounts of natural chemicals, emerging from defoliation and by the passing of tropical vegetation, are devoured by the soil animals so rapidly that the deterioration of this material is quicker than the rate at which it is supplanted.

High temperatures and high mugginess make conceivable a fast disintegration or mineralization of natural chemicals that the tropical soil is more unfortunate in nutrients than normal field soil in calm environments.

Plants in their indigenous habitats are constantly harmed by animals and by climate conditions like floods. They fill in water that for in any event, a piece of the developing time frame is sloppy and brimming with pollutants that choose the leaves and hinder the working of the plants. The lower part of tropical pools and streams is frequently peaty, and peat is obviously a helpless source of nutrients.

In the aquarium we can't cling to every one of these characteristic conditions in light of the fact that in a bound territory they are clearly negative to the plants. Rather we set up a vastly improved medium which has a significant number of the benefits of the regular habitat without its drawbacks.

Not at all like most land plants, aquatic plants are not reliant exclusively on sustenance got through the root framework. They can get nutrients through the entire surface of the plant body, particularly the epidermis of the leaves. Such commonly submerged plants as Ceratophyllum and Utricularia don't frame roots by any stretch of the imagination, while Elodea and Najas structure without a doubt, exceptionally short ones.

Hence the synthesis of the soil as well as of the water is vital, as the water should contain numerous mineral chemicals retained straight by the plant. This should be mulled over while setting up an aquarium and choosing the planting medium.

There has for quite some time been a debate about whether such established submersed plants as Cryptocoryne and Echinodorus get most of their nutrients through the leaves or the roots. Either answer is mostly right, as the technique for ingestion shifts with various centralizations of nutrients in the medium.

When in doubt, the more weakened the arrangement, the more available the minerals. On the off chance that there is a weighty centralization of nutrients in the soil, just the little rate broke up in the water will really be accessible to the plant.

Every so often, aquarists have best of luck developing plants for a while or more when utilizing such regularly questionable base materials as peat, garden soil, mechanical compost, or soil. This is on the grounds that we are never actually very sure about the appropriate ties of the substrate just from its name.

It is important to decide numerous components, like the pH, the extents of the significant nutrients (nitrates, phosphates, and so on), the presence or non-appearance of certain significant microelements, the measure of humus, and so on. Relatively few aquarists can test for this vital data without admittance to an exceptional research center. Business soil testing units are accessible for a portion of the substrate parts, yet not for all.

The most secure and most uniform strategy for setting up the planted aquarium is to get going with plain washed sand or rock and for the main year or so utilize just plants with solid and very much created rhizomes.

The rhizomes ought to have sufficient put away food stores to last the plant in any event a while. Great beginning plants incorporate the more grounded types of Echinodorus, Cryptocoryne, Lagenandra, Aponogeton, and Crinum. In a recently settled aquarium we can't develop such plants as Myriophyllum, Ludwigia, or Cabom-ba, as these needs save food supplies.

Start the aquarium by buying great quality rather coarse sand from a legitimate vendor. Ensure the sand is washed clean, everything being equal. On the lower part of the aquarium, introduce a decent quality under gravel channel to guarantee the course of the oxygenated water through the sand and around the foundations of the plants.

This rate the decay of natural chemicals, for example, fish dung and food stays to the mineral state and gives a steady temperature all through the aquarium. By utilizing a sand substrate along with an under-gravel channel, we can be certain that the water won't contain an overflow of natural chemicals and will be provided with adequate mineral nutrients important for the great improvement of the plants.

To acquire the greatest development and life span from your plants, the under-gravel channel ought to ideally be the sort comprising of various little punctured plastic cylinders. With this sort of channel, there is no hole between the foundation of the aquarium and the sand bed.

Accordingly, there is a bad situation for trash to develop where it can't be disintegrated by the appropriate microscopic organisms within the sight of oxygen. The cylinder type channel additionally has less possibility of getting tangled in the developing foundations of the plants.

For the first year, the substrate is poor in quite a while, yet with the progression of time, it gets more extravagant and more extravagant because of the collection and decay of fish and plant squanders. The sort of fish kept can somewhat decide how quick nutrients gather, as some fish are normally "dirtier" than others.

Consequently, a large portion of the live-carriers, including the swordtails, platies, mollies, and gup-pies, produce dirtier and more extravagant bottoms than the quadruplicate of comparative size. In the ordinary aquarium, there is a lot of rubbish, typically more than needed for the plants.

It should every so often be wiped out, as just the rubbish under the outside of the sand is truly important for the plants to acquire adequate nutrients. An aquarium a few years of age is an ideal climate for the development and improvement of all types of aquatic plants.

It is significant that the garbage in the sand ought to be decayed within the sight of oxygen. This deterioration, called vigorous, doesn't happen when the substrate medium is impermeable. Sealed shut media incorporate such things as peat, soil, and nursery soil.

Anaerobic deterioration (decay without air) brings about the arrangement of harmful gases like hydrogen sulfide. These gases show up as air pockets which hole out of the base when the plants are pulled out; the roots will be earthy colored or dark. Hydrogen sulfide can obliterate the two plants and fish. Modern composts are a source of hydrogen sulfide, so they ought not be utilized in the aquarium. Fish produce enough nutrients for the plants.

5.12. CONCLUSION

This chapter dealt with the ecology and management of aquatic plants. From describing it's various types like free-floating, water lettuce, emergent and submergent plants. It intends to define the ecology of aquatic plants by defining the macrophytes, it's classification along with depicting the challenges and adaptations to life in water. It also tells the different biotic interactions by illustrating the competition for light, allelopathy, herbivory and many other biotic interactions.

This chapter also states the management of aquatic plants along with describing the aquatic plant management program, schemes and strategies. It goes on to describe the long-term management like nutrient control, proper use of land, its planning and zoning, wise consumer use of commercial products and the treatment of inflowing waters.

It also emphasised on the short-term management of aquatic plants, the biological control, mechanical harvesting, environmental manipulation, and use of herbicides with the intend to elaborate control methods for aquatic plants in ponds and lakes.

There are various aquatic plant control methods which was portrayed and explained in the chapter, such as dredging and deepening, water level management, harvesting, shading and chemical dyes, pond bottom lining, biological controls, and chemical control. Towards the end of the chapter, there are general principles for the cultivation of aquatic plants which were explained.

REFERENCES

1. Bornette, G., & Puijalon, S., (2009). *Macrophytes: Ecology of Aquatic Plants. Encyclopedia of Life Sciences,* [online] Available at: https://www.researchgate.net/publication/229542529_Macrophytes_Ecology_of_Aquatic_Plants (accessed on 31 July 2021).
2. Carole A. Lembi, (2009). *Aquatic Plant Management: Identifying and Managing Aquatic Vegetation.* [ebook] Purdue University, pp. 1–19. Available at: https://www.extension.purdue.edu/extmedia/apm/apm_3_w.pdf (accessed on 31 July 2021).
3. Doug Crouch, (n.d.). *Aquatic Plant Ecology – Permaculture Design Course Handbook.* [online] Permaculture Design Course Handbook. Available at: https://treeyopermacultureedu.com/chapter-13-aquaculture/useful-aquatic-plants/ (accessed on 31 July 2021).
4. Doug Ludemann, (2020). *How to Grow Freshwater Aquarium Plants: 15 Steps* (with Pictures). [online] Wikihow.com. Available at: https://www.wikihow.com/Grow-Freshwater-Aquarium-Plants (accessed on 31 July 2021).
5. Fatimata Niang Diop, (2010). *Integration of freshwater biodiversity into africa's development process: mobilization of information and demonstration sites.* [ebook] Dakar-Fann, pp. 1–59. Available at: https://www.iucn.org/sites/dev/files/import/downloads/module_aquatic_plants_eng.pdf (accessed on 31 July 2021).
6. Ecology Center, (2021). *General Principles for the Cultivation of Aquatic Plants – Aquarium Plants.* [online] *Ecology Center.* Available at: https://www.ecologycenter.us/aquarium-plants/general-principles-for-the-cultivation-of-aquatic-plants.html (accessed on 31 July 2021).
7. Helfrich, L. A., Neves, R. J., Libey, G., & Newcomb, T. (2009). *Control Methods For Aquatic Plants in Ponds and Lakes.* [online].vt.edu. Available at: https://www.pubs.ext.vt.edu/420/420–251/420–251.html (accessed on 31 July 2021).
8. Water Resources Division, (2021). *Management of Aquatic Plants.* [ebook] Michigan: EGLE Environmental Assistance Center, pp. 1–19. Available at: https://www.michigan.gov/documents/deq/wrd-anc-AquaticPlantMgmt_408522_7.pdf (accessed on 31 July 2021).

Chapter 6

Organs and Their Functions in Water Plants

CONTENTS

This chapter deals with the organs and their functions in water plants, it begins with the plant body and plant tissues, and the plant organ systems, it explains the meristems and genetic control of flowers, the vegetative parts of plants, like the roots and structures.

It also throws light on the different types of roots, how roots grow and are used as food. It takes the reader to the terminology of stems, structures, different types of stems. It gives insight into the movement of water and minerals in the xylem, transportation of photosynthesis in phloem. It also goes on to explain the nutritional requirements of plants, its chemical composition, plant reproductive development and structure, sexual reproduction in gymnosperms and angiosperms.

6.1. THE PLANT BODY

Plants are made up of meristematic and permanent tissues and are supported by shoot and root organ systems.

6.1.1. Plant Tissues

Plants are multicellular eukaryotes with tissue systems made from various cell types that perform specific functions. There are mainly two types of plant tissue systems: meristematic tissue and permanent (or non-meristematic) tissue.

Cells from the meristematic tissue are found in meristems. Meristems are the part of the plant where continuous cell division and growth take place. Meristematic tissue cells can be both undifferentiated or incompletely differentiated regardless they divide and help the plant grow. On the other hand, cells that make up the permanent tissue of the plant do not divide actively.

On the basis of location, there are three types of Meristematic tissues in the plant. The meristematic tissue located at the tips of stem and roots are called Apical Meristems, they help a plant in growing lengthwise. The meristematic tissues which aid a plant to grow its girth or width are called Lateral Meristems.

The meristematic tissues that occur only in monocots at the base of leaf blades and at nodes (the area where leaves attach to a stem) are called Intercalary Meristems, and they help monocot leaf blades in growing lengthwise. This is what allows lawn grass to elongate even after regular mowing.

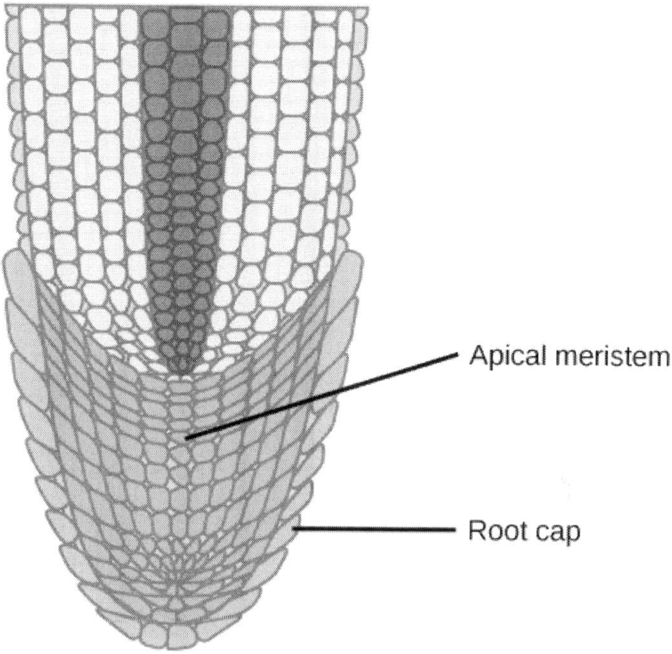

Figure 6.1: An illustration of apical meristem.

Source: Image by Wikimedia commons.

The cells produced by meristems differentiate quickly, or specialize and become permanent tissue. These cells perform specific tasks and lose the ability to divide further. They are classified into three main types: dermal, vascular, and ground tissue.

The dermal tissue is responsible for covering and protecting the plant. The vascular tissue is responsible for transporting water, minerals, and sugars to different parts of the plant. The Ground tissue serves as a site for photosynthesis, provides a supporting matrix for the vascular tissue, and helps to store water and sugars.

Plant tissues can be both simple (composed of similar cell types) or complex (composed of various cell types). The dermal tissue is a simple tissue covering the outer surface of the plant, it controls gas exchange. The vascular tissue is a complex one, made up of two tissues with specific roles: xylem and phloem.

Xylem transports water and nutrients from the roots to the other parts of the plant. It consists of three distinctive types of cells: vessel elements and tracheids (each of which conducts water) and xylem parenchyma. Phloem transports organic compounds from the site of photosynthesis to the other parts of the plant.

It is made up of four different types of cells: sieve cells (which conduct photosynthates), companion cells, phloem parenchyma, and phloem fibers. Phloem-conducting cells are alive at maturity, unlike xylem-conducting cells. The xylem and phloem are always observed to be lying adjacent to each other. In stems, the xylem and the phloem occur in the form of a vascular bundle; in roots, this is termed the vascular stele or vascular cylinder.

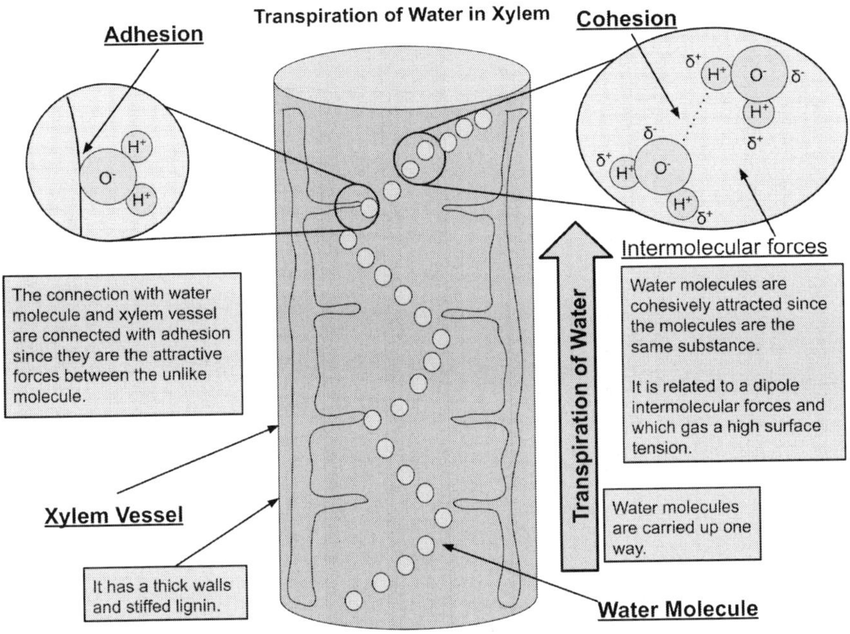

Figure 6.2: Transpiration of water molecule in xylem.

Source: Image by Wikimedia commons.

6.1.2. Plant Organ Systems

Just as in animals, in plants too, similar cells operating collectively form a tissue. Different types of tissues working together to perform one function, form an organ; organs working collectively form organ systems. There are two distinctive organ systems in a Vascular plant: a shoot system and a root system.

The shoot system is made up of two portions: the vegetative (non-reproductive) components of the plant, like the leaves and the stems; and the reproductive components of the plant, i.e., flowers, and fruits. The shoot system typically grows above ground, where it absorbs the light needed for photosynthesis. The root system, which supports the plants and absorbs water and minerals, is typically underground.

6.2. MERISTEMS

Plant meristematic tissues are cells that divide so as to create the numerous organs of a plant and keep it growing.

The adult body of vascular plants is the end result of meristematic activity. Plant meristems are made up of a group of undifferentiated self-renewing stem cells from which maximum plant structures arise; they are centers of mitotic cell division. Meristematic cells are also responsible for maintaining the growth of the plant.

The Shoot Apical Meristem (SAM) gives rise to organs like leaves and flowers, whereas the Root Apical Meristem (RAM) supports the meristematic cells in future root growth. Shoot and root apical meristem cells divide rapidly and are said to be indeterminate; this means that their future is uncertain. Meristematic cells are often compared to the stem cells in animals as they behave and function similarly.

6.2.1. Meristem Tissue and Plant Development

Meristematic tissues are cells or groups of cells that have the power to divide. These tissues in a plant consist of small, tightly packed cells that can keep dividing to create new cells. The Meristematic tissue is characterized by small cells, thin cell walls, large cell nuclei, absent or small vacuoles, and no intercellular spaces.

Meristematic tissues are found in several locations, including near the tips of roots and stems (apical meristems), in the buds and nodes of stems, in the cambium between the xylem and phloem in dicotyledonous trees and shrubs, beneath the epidermis of dicotyledonous trees and shrubs (cork cambium), and in the pericycle of roots, producing branch roots. The two kinds of meristems are primary meristems and secondary meristems.

6.2.2. Meristem Zones

The apical meristem, also known as the "growing tip," is an undifferentiated meristematic tissue located on the inside of buds and growing tips of roots in plants. Its primary function is to set off the growth of new cells in younger seedlings at the tips of roots and shoots and forming buds.

Apical meristems are divided into four zones:

1. the central zone,
2. the peripheral region,
3. the medullary meristem, and
4. the medullary tissue.

The central zone is found at the meristem summit, wherein a small group of slowly dividing cells can be observed. Cells of this zone have a stem cell function and are vital for meristem maintenance. The proliferation and growth rates on the meristem summit typically fluctuate extensively from the ones on the periphery. Surrounding the central zone is the peripheral zone.

The rate of cell division is higher in the peripheral zone as compared to that of the central zone. Peripheral zone cells give rise to cells that contribute to the organs of the plant, which include leaves, inflorescence meristems, and floral meristems.

An active apical meristem lays down a growing root or shoot behind itself, pushing itself forward. They are very small in comparison to the cylinder-shaped lateral meristems and are made up of numerous layers, which vary according to the type of plant. The outermost layer is known as the tunica, whilst the innermost layers are cumulatively known as the corpus.

6.3. GENETIC CONTROL OF FLOWERS

As proven through the ABC model, a variety of genes control flower development, which includes sexual maturation and growth of reproductive organs.

The process through which angiosperms produce a pattern of gene expression in meristems that causes the appearance of a flower is called flower development. A flower (also known as a bloom or blossom) is the reproductive structure observed in flowering plants. There are three physiological developments that should arise so that reproduction can occur:

1. the plant should have passed from sexual immaturity into a sexually mature state.
2. the apical meristem should have transformed from a vegetative meristem into a floral meristem or inflorescence.
3. the flowers individual organs should have grown (modeled using the ABC model).

6.3.1. Flower Development

A flower develops on a modified shoot or axis from a determinate apical meristem (determinate meaning the axis grows to a fixed size). The transition to flowering is one of the fundamental phase changes that a plant makes in the course of its life cycle.

The transition needs to take place at a time that is suitable for fertilization and the generation of seeds, thereby ensuring maximal reproductive success. To flower at a suitable time, a plant can understand vital endogenous and environmental cues like changes in levels of plant hormones and seasonable temperature, and photoperiod changes. Many perennials and most biennial plants require vernalization to flower.

6.3.2. Genetic Control of Flower Development

When plants sense a possibility to flower, signals are transmitted via florigen, and this involves a number of genes, which includes constans, flowering locus C, and flowering locus T. leaves produce florigen in reproductively favorable conditions, it acts in buds and growing tips to set off a number of specific physiological and morphological modifications.

From a genetic perspective, there are two phenotypic changes programmed in a plant that manipulate vegetative and floral growth. The first genetic change is the switch from the vegetative to the floral state. Flowering does not occur if this genetic change does not function properly.

The second genetic event follows the commitment of the plant to form flowers. The sequential development of plant organs points towards the existence of a genetic mechanism wherein a chain of genes is sequentially turned on and off. This switching is important for every whorl to achieve its final individual identity.

6.3.3. ABC Model of Flower Development

According to the simple ABC model of floral development, three gene activities (termed A, B, and C-functions) engage to decide the developmental identities of the organ primordia (singular: primordium) in the floral meristem.

The ABC model of flower development was developed in order to explain the collection of genetic mechanisms that establish floral organ identity in the Rosids and the Asterids. Each species has four verticils (sepals, petals, stamens, and carpels), which are defined by the differential expression of several homeotic genes found in every verticil.

Only A-genes are expressed in the first whorl, which results in the formation of sepals. Both A- and B-genes are expressed in the second whorl, which results in the formation of petals. In the third whorl, B and C genes interact to form stamens, and in the center of the flower C-genes alone form carpels.

For instance, if there is a lack of B-gene function, mutant flowers are produced with sepals in the first whorl as usual but also in the second whorl as opposed to the regular process of petal formation. In the third whorl, the lack of B function but the presence of C-feature mimics the fourth whorl, leading to carpels' formation.

The genes central in this model majorly belong to the MADS-box genes. They are transcription elements that alter the expression of the genes unique to every floral organ.

6.4. VEGETATIVE PLANT PARTS

External plant structures like leaves, stems, roots, flowers, fruits, and seeds are referred to as plant organs. Every organ is an organized group of tissues that works collectively to carry out a particular function. These structures can be divided into groups: sexually reproductive and vegetative.

Vegetative parts consist of roots, stems, shoot buds, and leaves; they're not involved in sexual reproduction directly. Vegetative parts are regularly utilized in asexual forms of reproduction like cuttings, budding, or grafting.

6.4.1. Roots

Roots are overlooked many a time, possibly due to the fact they are seen much less than the rest of the plant. However, it's essential to understand plant root systems since they have a significant contribution to a plant's size and vigor, method of propagation, adaptation to soil types, and reaction to cultural practices and irrigation.

Roots generally originate from the lower part of a plant or cutting. They have a root cap; however, they lack nodes and never bear leaves or flowers directly. Their predominant functions are to take in nutrients and moisture, hold the plant in the soil, support the stem, and store food. In a few plants, they are used for propagation as well.

6.4.2. Structure

Internally, there are three primary parts of a root:

1. On the tip is the meristem that manufactures new cells; this is the area where cell division and growth occur.
2. Located behind the meristem is the zone of elongation; this is the area where cells increase in size through food and water absorption. They push the root through the soil as they grow.
3. Directly underneath the stem is the zone of maturation. This is where cells develop into specific tissues such as the epidermis, cortex, or vascular tissue.

The outermost layer of cells of a root is known as its epidermis. These cells absorb water and minerals dissolved in water. Cortex cells are involved in transporting water from the epidermis to the vascular tissue (xylem and phloem) and in storing food. Vascular tissue is positioned in the middle of the root and transports food and water.

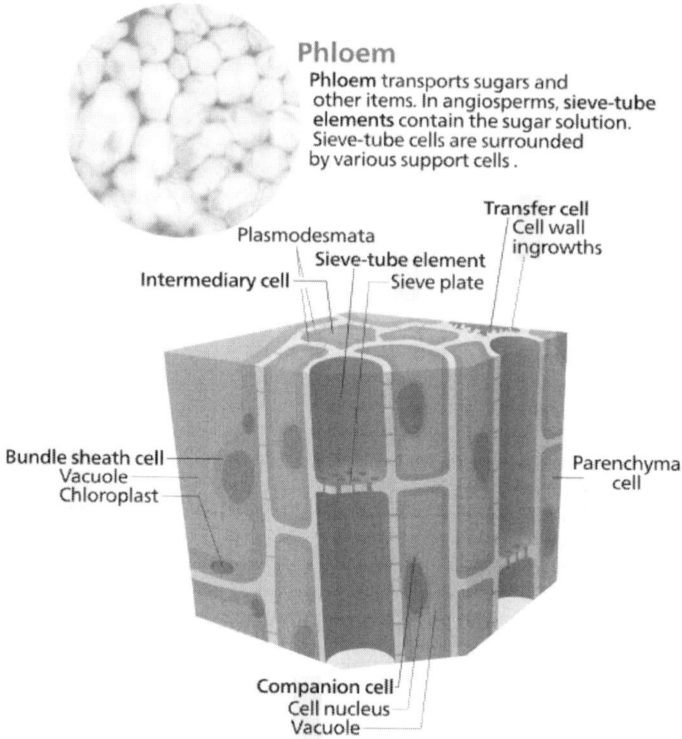

Figure 6.3: An illustration of phloem.

Source: Image by Wikimedia Commons.

Externally, there are two regions of importance: the root cap and the root hairs. The root cap is the root's outermost tip. It is made up of cells that the root sheds as it grows through the soil. It is responsible for guarding the root meristem.

Root hairs are delicate, elongated epidermal cells that are found in a small area right behind the roots growing tip. They appear to be very fine to the naked eye. They increase the root's surface area and absorptive capacity. Root hairs commonly stay alive for 1 or 2 days. They shed easily when a plant is transplanted or dry out in the sun.

Many roots have a naturally occurring symbiotic (mutually beneficial) relationship with certain fungi, which improves the plant's ability to take in water and nutrients. This beneficial association is referred to as mycorrhizae (fungus + root).

6.5. TYPES OF ROOTS

Roots have been broadly categorized into two types: primary and lateral. A primary root stems at the lower end of a seedlings embryo. If the primary root becomes the main feature of the root system, continues to elongate downward, and has restricted secondary branching, it is known as a taproot. Hickory and pecan bushes, in addition to carrots, have taproots.

A side or branch root that arises from another root is called a lateral, or secondary, root. If the primary root stops elongating and several lateral roots develop, a fibrous root system is formed. These lateral roots branch again and again to form the network of feeding roots found on most plants.

Fibrous root systems are produced naturally in some plants like grasses. In other cases, severing a plant's taproot by undercutting it may encourage the plant to produce a fibrous root system. This approach is used by nurseries with trees that naturally produce a taproot because trees with a compact, fibrous root system are transplanted much more successfully.

6.5.1. How Roots Grow

In the primary stages of development, a seedling absorbs nutrients and moisture from the soil around the sprouting seed. A layer of fertilizer several inches on every side and slightly under the newly planted seeds facilitates the early growth of most row crops.

As a plant establishes itself, the amount and distribution of its roots greatly affect its capacity to soak up moisture and nutrients. Most of the absorbing (feeder) roots are located in the top 12 inches of soil for maximum plants. The soil environment in this area is normally best suited for root growth, with a good balance of fertility, moisture, and air spaces.

The following factors are important in root growth:

1. Roots do not develop properly in water-saturated soil and may ultimately die because of lack of oxygen.
2. Roots grow much deeper in loose, well-drained soil as compared to heavy, poorly drained soil.
3. A dense, compacted soil layer can limit or terminate root growth.
4. Container plants not only have limited space for root growth but are also prone to cold damage as the restricted amount of soil surrounding their roots might not offer adequate insulation.

5. Roots don't just grow downwards; they also grow laterally and regularly extend well past a plant's dripline. This extensive root system should be kept in mind while disturbing the soil around any trees or shrubs.

6.5.2. Roots as Food

An enlarged root is the edible part of many vegetable crops. Sweet potatoes are a swollen tuberous root; and carrots, parsnips, salsify, and radishes are elongated taproots.

6.6. STEMS

The stem supports the buds and leaves and acts as a conductor for the transportation of water, minerals, and food (photosynthates). The vascular system in the stem forms a continuous path from the root to stem to leaf, it is through this system from which water and food flow.

6.6.1. Stem Terminology

Shoot: A young stem (1-year-old or less) with leaves.

Twig: A young stem (1-year-old or less) that is in the dormant winter stage (has no leaves).

Branch: A stem that is more than 1 year old, typically with lateral stems radiating from it.

Trunk: A woody plant's main stem.

6.7. STRUCTURE

6.7.1. Vascular System

This system consists of the xylem, phloem, and vascular cambium. Think of it as the plumbing system of a plant. Xylem tubes transport water and dissolved minerals. Phloem tubes transport food in the form of sugar. The Cambium is a layer of meristematic tissue that separates the xylem and phloem and continuously produces new xylem and phloem cells. This new tissue is responsible for increasing the circumference of the stem.

The vascular cambium is very important to gardeners. For example, the tissue of grafted scion and rootstock must be coordinated. In addition, careless pruning of weeds can tear off the bark, damage the cambium, and cause the tree to die. The vascular system of monocots and dicots is different. Although both contain xylem and phloem, the location of these structures in each structure is different. In monocots, the xylem and phloem are combined into bundles and distributed along the stem.

In dicots, the vascular system is called continuous because it forms rings in the stem. The phloem ring is located near the bark and ultimately becomes a part of the bark in mature woody stems. The xylem forms the inner ring. In woody plants, this is called sapwood and heartwood. The difference in the vascular system of monocots and dicots is of practical interest to gardeners because certain herbicides only act on one group. For example, 2,4-D only kills plants that have a continuous vascular system (dicots). On the other hand, no matter what the vascular system of the plant is, non-selective herbicides (such as glyphosate) will kill the plant.

6.7.2. Nodes

The node is the part of the stem on which the bud is. This is a place full of vitality and cell growth, where small buds grow into leaves, stems, or flowers. It is important to find the nodes of a plant when pruning. Usually, you want to trim near, but not too close to a node. Pruning encourages buds at that node to develop and eventually form new stems or leaves.

The area between two nodes is known as an internode. Its length depends on many factors, including genetic factors. Various other factors also affect the length of the internodes:

1. A decrease in soil fertility will reduce the length of the internodes, and the use of high-nitrogen fertilizer will significantly increase it.
2. Not getting sufficient light increases internode length and leads to thin stems. This condition is called stretch or etiolation, and it is common in seedlings grown indoors and indoor plants that do not receive enough sunlight.
3. The length of the internode also changes with the season. The internodes grown in the early season are longer, and those grown in the late season are usually much shorter.
4. If a stem's energy is divided among three or four side stems or is diverted into fruit growth and development, internode length is shortened.

5. Herbicides and other plant growth regulator substances can also affect the length of the internodes.

6.8. TYPES OF STEMS

Stems can be very long, with large distances between the leaves and buds (e.g., branches of trees, runners on strawberries) or they can be compressed, and there can be smaller spaces between the buds or leaves (e.g., strawberry crowns, fruit spurs, and African Violets). Although stems usually grow above the ground, they sometimes grow underground in the form of rhizomes, tubers, bulbs, or corms. All stems must have buds or leaves to be classified as stem tissue.

6.8.1. Specialized Aboveground Stems

Some plants have special aboveground stems called crowns, spurs, or stems. The crown (found in strawberries, dandelions, and African violets) is a compressed stem with leaves and flowers on short internodes. The short and stubby side stems that arise from the main stem are called spurs.

The spurs of pears, apple trees, and cherries are the ones that bear fruits. Severe pruning near the fruit-bearing spurs can make them stop bearing fruits, terminating the possibility of harvest that year.

Long and fleshy or semi-woody horizontal stems usually found on the surface of the soil are known as stolons. Strawberry runners are stolons with small leaves on the nodes. From these nodes, the roots develop and form daughter plants. This vegetative reproduction is an easy way to expand a strawberry field. Spider plants also produce stolons, which can eventually grow into new plants.

6.8.2. Specialized Below-Ground Stems

Potato tubers, iris rhizomes, and tulip bulbs are underground stems that store food for plants. Sometimes it is difficult to tell the root and stem apart, but the safest way to be sure is to look for nodes as they only occur in stems and not in roots.

For example, in potato tubers, the "eyes" are actually stem nodes, and each eye contains a set of buds. When planting potatoes from seeds, it is very important that each piece has at least one eye, approximately the size of a golf ball, so as to provide enough energy for the early growth of shoots and roots. Rhizomes look a lot like stolons because they grow horizontally between plants. Some rhizomes are compressed and fleshy (such as iris), while others are thin and have elongated internodes (such as bentgrass). Johnsongrass is an insidious weed, mainly due to the spreading capability of its rhizomes.

Figure 6.4: An illustration of bentgrass.

Source: Image by Flickr.

Tulips, lilies, daffodils, and onions produce bulbs, which are compressed and shortened underground stems surrounded by fleshy scales (leaves) that wrap the central bud at the tip of the stem. In November, you can cut the bulb of a tulip or daffodil in half and view a miniature of all parts of the flower.

After a bulb-bearing plant blooms, its phloem transfers food reserves from the leaves to the scales of the bulb. It uses stored food to grow in the spring. It is important not to remove the leaves of daffodils, tulips, and other bulb-producing plants until they turn yellow and dry. After that, they stop producing food that would be used for flowering next year.

The two types of bulbs are tunicate and non-tunicate. Tunicate bulbs (such as daffodils, tulips, and onions) have a thin paper-like coating which are actually modified leaves. It prevents the bulb from any harm that might come from digging and drying up once it is out of the soil. Nontunicate bulbs (such as lilies) have no such paper-like coating. They are very easy to damage and easily dry out, thus they need to be handled very carefully.

Corms are another type of underground stem. Although both bulbs and corms are made of stem tissue, they are not the same. Corms are shaped like bulbs but have no fleshy scales. A corm is a hard, swollen stem with dry, scaly leaves. Gladiolus and crocuses produce corms.

Some plants like Begonias and cyclamen form a modified underground stem called a tuberous stem. These stems are short, flat, and enlarged. Buds and shoots come out from the top half (crown), while fibrous roots grow from the bottom half.

Other plants (such as dahlias and sweet potatoes) produce underground storage organs called tuberous roots, which are often mistaken for bulbs and tubers. However, these are root tissue, not stem tissue, and have no nodes or internodes.

6.8.3. Stems and Propagation

Stems are usually used for the vegetative propagation of plants. For many ornamental plants, the use of aboveground stems containing nodes and internodes is an effective method of propagation. These cuttings lead to root formation and eventually new plants.

Below-ground stems are also good propagative tissues. You can divide the rhizome into several parts; remove small bulblets or cornels from the mother plant; cut the tuber into small pieces with eyes and nodes. All these tissues will produce new plants.

6.9. MOVEMENT OF WATER AND MINERALS IN THE XYLEM

Transpiration helps to move water and minerals in the xylem, but it needs to be controlled to prevent water loss.

Most plants get the water and minerals they need from the roots. The path chosen: soil->root->stem->leaves. Minerals (such as K^+, Ca^{2+}) dissolve in water and then travel (usually accompanied by various organic molecules supplied by root cells). Water and minerals enter the roots in different ways and eventually converge at the stele or central vascular bundle at the roots.

Transpiration is the loss of water by plants due to evaporation from the surface of the leaves. This is the main driving force for water movement in the xylem. Transpiration is caused by the evaporation of water on the leaves or at the atmospheric interface; negative pressure (tension) is generated on the surface of the leaves, which is equivalent to –2 MPa.

This value varies greatly and is dependent on the vapor pressure deficit, it can be insignificant at high relative humidity (RH) and substantial at low relative humidity. Due to this tension, the water rises from the roots.

When the stomata are closed at night and transpiration stops, the water in the stems and leaves is blocked due to the adhesion of water molecules to each other and the adhesion of water to the cell walls of the xylem vessels and tracheid. This is called the cohesion-tension theory of sap ascent.

The cohesion-tension theory tells us how water moves upward through the xylem. At a cellular level inside the leaf, Cellulose microfibrils are saturated by the water on the surface of the primary cell wall. The leaves contain many large intercellular air spaces, which are used to exchange oxygen for carbon dioxide, which is essential for photosynthesis.

The moist cell wall is exposed to the internal air space, and water from the cell surface evaporates into the air space. This reduces the film on the surface of the mesophyll cells. The reduction leads to an increase in the tension on the water in the mesophyll cells, hence increasing the pull of water in the xylem vessels. In a xylem container.

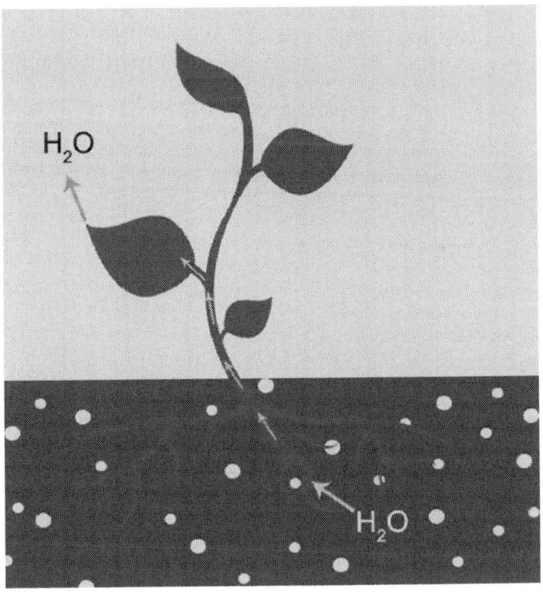

Figure 6.5: Transpiration in plants.

Source: Image by Flickr.

The xylem vessels and tracheids are structurally adapted to large changes in pressure. The small holes between the components of the vessel reduce the amount and size of gas bubbles formed by cavitation. The formation of air bubbles in the xylem is disadvantageous because it interrupts the continuous flow of water from the root to the top of the plant and causes the flow of xylem sap to be interrupted (embolism).

The taller the tree, the greater the pulling force required to pull water in a continuous column, which increases the number of cavitation opportunities. In larger trees, these embolisms can block the xylem vessels, making it impossible for them to operate.

6.9.1. Control of Transpiration

Transpiration is a passive process: it does not require metabolic energy in the form of ATP to move water. The energy that drives transpiration is the energy difference between the water in the soil and the water in the atmosphere. However, transportation is strictly controlled. Exposure of leaves to the atmosphere causes transpiration, but it can also cause a large loss of water. Up to 90% of the water absorbed by the roots can be lost through transpiration.

The leaves are covered with a waxy cuticle to prevent water loss. The regulation of transpiration is mainly achieved by opening and closing the stomata on the surface of the leaf. The stomata are surrounded by two specialized cells called guard cells, these cells open and close according to environmental signals such as the intensity and quality of light, the state of water in the leaves, and the concentration of carbon dioxide.

Stomata need to be opened for carbon dioxide-containing air to flow through so that it can diffuse with oxygen into the leaves and allow photosynthesis and respiration to occur. However, when the pores are opened, water vapor will be lost to the external environment, thereby increasing the rate of transpiration. Therefore, plants need to maintain a balance between effective photosynthesis and water loss.

Over time, plants have evolved to adapt to their surroundings and reduce transpiration. Desert plants (xerophytes) and plants that grow on other plants (epiphytes) can only get limited water. The waxy cuticle of these plants is usually much thicker than the ones that grow in a mild, well-hydrated environment (mesophytic plants). Aquatic plants (hydrophytes) have their own set of anatomical and morphological adaptations to leaves.

Usually, xerophytes and epiphytes have a thick layer of trichomes or stomata sunken below the leaf surface. Trichomes are special, hair-like epidermal cells that secrete oil and other substances. These adaptations impede the flow of air through the stomata and reduce transpiration. Many times, multiple epidermal layers are found in such plants.

6.10. TRANSPORTATION OF PHOTOSYNTHATES IN THE PHLOEM

Translocation moves photosynthates from the source to the sink through the phloem.

Plants need a source of energy to grow. In seeds and bulbs, food is stored in polymers (such as starch), which are metabolized into sucrose for the development of new plants. Plants can produce their own food by photosynthesis after shoots and leaves begin to grow. Products of photosynthesis (called photosynthates) usually exist in the form of simple sugars such as sucrose.

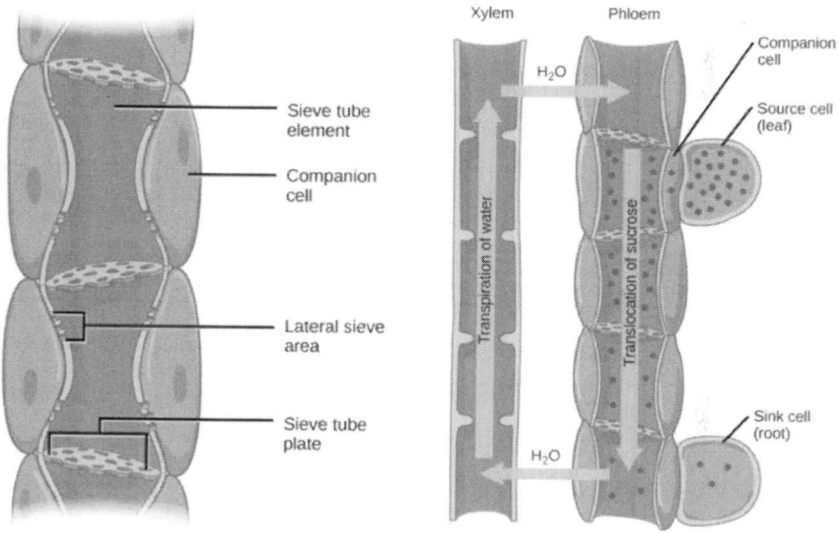

Figure 6.6: Transportation of photosynthates in the phloem.

Source: Image by Lumen Learning.

6.10.1. Sources and Sinks

The source is the structure that produces photosynthates in growing plants. Sugar from sources such as leaves must be transported to the growing parts of the plant. These sugars are transported in the plant through the phloem, and this is called translocation.

The spots where sugar needs to be delivered such as roots, young shoots, and developing seeds are called sinks. Seeds, tubers, and bulbs can be the source or the sink, depending on the growth stage of the plant and the time of year. Usually, the products from the source are translocated to the closest sink via the phloem. For example, the photosynthates from the upper leaves travel to the top of the growing shoot tip, and the photosynthates in the lower leaves travel to the roots.

Intermediate leaves send their products in both directions. The flow of substances in phloem is multidirectional; on the other hand, the flow in the xylem is always unidirectional (from the ground to the leaves to the atmosphere).

As the plant grows and develops the pattern of photosynthate flow changes. Photosynthates mainly affect the roots in the early stage of development, the shoots and leaves in the vegetative growth period, and the seeds and fruits in the reproductive development period. They are also directed at the tubers for storage.

6.10.2. Translocation: Transport from Source to Sink

Photosynthates are produced in the mesophyll cells of the leaves undergoing photosynthesis. They are then translocated through the phloem to where they are used or stored. Mesophyll cells are connected by cytoplasmic channels called plasmodesmata.

Photosynthates pass through the plasmodesmata to reach the phloem sieve tube elements (STEs) in the vascular bundles. The photosynthates from the mesophyll cells are loaded into the STEs of the phloem. Using the electrochemical potential of the proton gradient, sucrose is actively transported into the phloem cells against its concentration gradient (a process that requires ATP). It is combined with the absorption of sucrose by a carrier protein called sucrose-H + symporter.

Phloem STEs have low cytoplasm content and are connected by sieve plates with pores so that the phloem sap can flow or be translocated by pressure-driven bulk flow. Companion cells are associated with STEs. They contribute to metabolic activity and generate energy for STEs.

In the phloem, photosynthates are translocated to the nearest sink. Phloem sap is an aqueous solution that contains up to 30% sugars, minerals, amino acids, and plant growth regulators. High sugar content will reduce Ψs, thereby reducing total water content, causing water to move out of the adjacent xylem to the phloem tubes.

The flow of water increases the water pressure in the phloem, causing a large amount of phloem sap to flow from the source to the sink. In the sink cells concentration of sucrose is lower than in the phloem STEs as the sink sucrose has been metabolized for growth or converted to starch (for storage) or other polymers (for structural integrity).

At the sink end of the phloem tube, unloading occurs by either diffusion or active transport of sucrose molecules from an area of high concentration to one of low concentration. Water diffuses out of the phloem by osmosis, and then transpires or is recycled back to phloem sap through the xylem.

6.11. NUTRITIONAL REQUIREMENTS OF PLANTS

In addition to absorbing the necessary sunlight, plants also absorb nutrients in the soil, water, and carbon dioxide to meet their nutrient needs for growth.

Plants are unique organisms that can absorb nutrients and water through their roots, as well as carbon dioxide from the atmosphere. Quality of soil and climate are the main factors for plant distribution and growth. Nutrients from the soil, water, and carbon dioxide work together with sunlight to enable plants to grow. In order for plants to be mature and fertile, many requirements must be met, and many events must be coordinated.

For example, the Cucurbitaceae plant family was originally cultivated in Mesoamerica, although there are several species that are native to North America. They include many edible species such as squash and pumpkin as well as inedible gourds.

First, the seeds must germinate in the soil under the right conditions. Therefore, temperature, humidity, and soil quality are important factors affecting the germination and development of seedlings. Soil quality and climate are critical to the spread and growth of plants.

Secondly, young seedlings will eventually become mature plants, and their roots will absorb nutrients and water from the soil. At the same time, the parts of the plant above the ground absorb carbon dioxide from the atmosphere and use the energy from sunlight to produce organic compounds through photosynthesis. Eventually, the fruit grows and matures, and the cycle begins with new seeds.

There is a complex dynamic relationship between plants and soil, which ultimately decides the results and viability of plants. In the following sections of this chapter, we will discuss many aspects of plant nutritional requirements in more detail.

6.12. THE CHEMICAL COMPOSITION OF PLANTS

Plants are made up of water, carbon-containing organics, and non-carbon-containing inorganic substances such as potassium and nitrogen.

6.12.1. Water

Water absorption by the roots: Water is absorbed by root hairs and rises through the xylem to the leaves.

Since plants need nutrients in the form of elements such as carbon and potassium, it is crucial to understand the chemical composition of plants. The volume of a plant cell is majorly water; it usually accounts for 80% to 90% of the total weight of the plant.

Soil is the source of water for land plants. Even if it looks dry, it may be a rich source of water. The roots of plants absorb water from the soil through root hairs and transport it upwards to the leaves through the xylem.

When the water vapor leaves the leaves, the transpiration process and the polarity of the water molecules (which allows them to form hydrogen bonds) draw more water from the roots of the plant to the leaves. Plants need water to maintain cell structure, metabolic function, transport nutrients and perform photosynthesis.

6.12.2. Nutrients

In order to sustain life, plant cells need essential substances, which are collectively referred to as nutrients. Plant nutrients can be composed of organic or inorganic compounds. Organic compounds are compounds containing carbon extracted from the atmosphere, such as carbon dioxide.

Carbon from the atmospheric CO_2 makes up most of the dry matter of most plants. Inorganic compounds do not contain carbon and are not part of or produced by living organisms. Inorganic substances (which make up most of the soil) are often called minerals: the ones needed by plants include nitrogen (N) and potassium (K) for structure and regulation.

6.12.3. Essential Nutrients for Plants

Approximately 20 macronutrients and micronutrients are considered essential nutrients to aid all the biochemical needs of plants.

Plants need only light, water, and about 20 elements to support all their biochemical needs. These 20 elements are called essential nutrients. For an element to be regarded as essential, three criteria are required:

1. Without the element, the plant cannot complete its life cycle;
2. No other element can perform the function of the element;
3. This element is directly involved in plant nutrition.

6.12.4. Macronutrients and Micronutrients

The main elements can be divided into macronutrients and micronutrients. Nutrients that are required by plants in larger quantities are called macronutrients. About half of the main elements are considered macronutrients: carbon, hydrogen, oxygen, nitrogen, phosphorus, potassium, calcium, magnesium, and sulfur.

The first of these macronutrients is carbon (C), which is used to form carbohydrates, proteins, nucleic acids, and many other compounds. Therefore, it exists in all macromolecules. On average, the dry weight of cells (excluding water) is 50% carbon, which makes it an important part of plant biomolecules.

Essential elements required by plants: If without an element, a plant cannot complete its life cycle, no other element can replace it, and the element directly participates in the nutrition of the plant, then it is regarded as an essential.

The second most abundant element in plant cells is nitrogen (N). It is part of proteins and nucleic acids. Nitrogen is also used in the synthesis of certain vitamins. Hydrogen and oxygen are macronutrients present in a large number of organic compounds, and they also form water. Oxygen is important for cellular respiration.

Plants use oxygen to store energy in the form of ATP. Phosphorus (P) is another macromolecule, it is necessary for the synthesis of nucleic acids and phospholipids. As part of ATP, phosphorus enables food to be converted into chemical energy through oxidative phosphorylation. In the process of photophosphorylation, light energy is converted into chemical energy in photosynthesis. And the chemical energy extracted during breathing. Sulfur is part of some amino acids such as cysteine and methionine and is present in several coenzymes. Sulfur also plays a role in photosynthesis as part of the electron transport chain, where the hydrogen gradient plays a key role in the conversion of light energy into ATP.

Potassium (K) plays an important role in regulating the opening and closing of stomata. As an opening for gas exchange, the stomata help maintain a healthy water balance. The potassium ion pump supports this process. Magnesium (Mg) and calcium (Ca) are also important macronutrients. Calcium has a dual role: it regulates the transportation of nutrients and supports many enzyme functions. Magnesium is essential for photosynthesis. These minerals and micronutrients also contribute to the ion balance of plants.

In addition to macronutrients, organisms also need small amounts of various elements. The content of these trace elements or micronutrients is very small. The seven essential microelements are boron, chlorine, manganese, iron, zinc, copper, and molybdenum. It is believed that boron (B) is involved in the transport of carbohydrates in plants.

It also helps regulate metabolism. Boron deficiency usually causes bud dieback. Chlorine (Cl) is important for osmosis and ion balance. It also plays a role in photosynthesis. Copper (Cu) is found in certain enzymes. Symptoms of copper deficiency include browning and chlorosis (yellowing) of leaf tips. Iron (Fe) is necessary for the synthesis of chlorophyll; hence iron deficiency causes chlorosis. Manganese (Mn) activates several important enzymes involved in the formation of chlorophyll. P

lants lacking manganese develop chlorosis between the veins. The availability of manganese depends in part on the pH of the soil. Molybdenum is very important for plant health as plants use it to reduce nitrates into usable forms.

Some plants use it for nitrogen fixation. Therefore, it may be necessary to add it to the soil before planting legumes. Zinc (Zn) is involved in the formation of chlorophyll and also activates many enzymes. Symptoms of zinc deficiency include chlorosis and stunted growth.

The lack of any of these nutrients (especially macronutrients) will adversely affect the growth of plants. Deficiency can cause stunted growth, slow growth, or chlorosis, depending on the nutrients. Extreme deficiency can cause the leaves to show signs of cell death.

6.13. PLANT REPRODUCTIVE DEVELOPMENT AND STRUCTURE

Plants have advanced diverse sexual and asexual techniques to ensure reproductive success.

Plants have developed distinctive reproductive techniques for the continuation of their species. Some plants reproduce sexually whilst others reproduce asexually, animals on the other hand, depend nearly completely on sexual reproduction.

Plant sexual reproduction generally relies upon pollinating agents, whereas asexual reproduction does not require these agents. Flowers are usually the showiest or most strongly fragranced part of plants. With their vibrant colors, fragrances, and interesting shapes and sizes, flowers appeal to insects, birds, and animals to serve their pollination needs. Other plants pollinate via wind or water; while some others self-pollinate.

6.13.1. Asexual Reproduction

Vegetative reproduction is a form of asexual reproduction. It is also known as vegetative propagation, clonal growth, or vegetative multiplication. Vegetative growth is the growth of the specific plant, whilst vegetative reproduction is any method whose outcome is a new plant, without the production of seeds or spores.

This process occurs naturally in many species and is also a method applied or endorsed by horticulturists and farmers to gain surplus quantities of economically feasible flowers. Thus, it could be said that it is a form of cloning that has been performed by humans for thousands of years and by plants for hundreds of millions of years.

6.13.2. Sexual Reproduction and the Flower

The reproductive organ of angiosperm plants is called the flower. All plants have the means and corresponding structures for reproducing sexually. The fundamental characteristic of a flower is to provide seeds through sexual reproduction.

Seeds are the following generation; they are the primary approach in maximum plants with the aid of using which individuals of the species are dispersed throughout the landscape. Actual dispersal is, in maximum species, a feature of the fruit (a structural component that commonly surrounds the seed).

6.14. SEXUAL REPRODUCTION IN GYMNOSPERMS

Gymnosperms produce male and female gametophytes in separate cones and rely on wind for pollination.

Like angiosperms, the life cycle of gymnosperms is also characterized by generational changes. In coniferous trees such as pine, the green part of the plant is the sporophyte; the cones contain both male and female gametophytes.

The female cone is larger than the male cone and is closer to the top of the tree, whereas the smaller male cones are located in the lower region of the tree. Since pollen is carried by the wind, this arrangement makes it hard for gymnosperms to self-pollinate as pollen is shed and blown by the wind.

6.14.1. Male Gametophyte

A male cone has a central axis on which bracts, a type of modified leaf, are attached. The bracts, known as microsporophyll, are the sites where microspores will develop. The microspores develop inside the microsporangium. Within the microsporangium, cells known as microsporocytes divide by meiosis to produce four haploid microspores.

Further mitosis of the microspore produces two nuclei: the generative nucleus and the tube nucleus. Upon maturity, the male gametophyte (pollen) is released from the male cones and is carried by the wind to land on female cones.

6.14.2. Female Gametophyte

The female cone also has a central axis on which bracts, called megasporophylls, are found. In female cones, there are megaspore mother cells found in the megasporangium. Megaspore mother cells divide through meiosis to form four haploid megaspores.

One of the megaspores divides to form the multicellular female gametophyte, while the other megaspores divide to form the rest of the structure. Female gametophytes are contained in structures called archegonium.

6.14.3. Reproductive Process

Tube cells of the pollen land in the female cone and form pollen tubes, through which generative cells migrate to the female gametophyte through the micropyle. The pollen tube grows and migrates to the female gametophyte in about one year.

The male gametophyte containing the generative cell, which is divided into two sperm nuclei, one fused with the egg cell, and the other degenerates.

After the egg is fertilized, a diploid zygote is formed, which forms an embryo through mitosis.

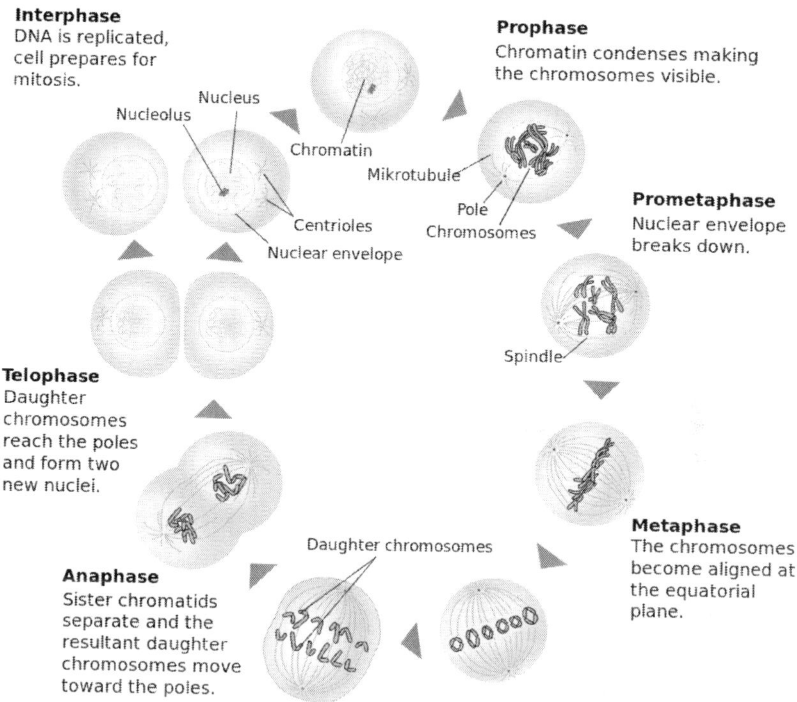

Figure 6.7: Process of mitosis in plants.

Source: Image by Wikimedia Commons.

During seed development, the scales of the cones close. The seeds are covered with seed coats obtained from female sporophytes. It takes one to two years for the seeds to develop. Once the seeds are ready to disperse, the bracts of the female cones open up, allowing the seeds to spread; because gymnosperm seeds are not coated, they will not bear fruit.

6.14.4. Angiosperms versus Gymnosperms

Gymnosperm reproduction differs from that of angiosperms in numerous ways. In angiosperms, the female gametophyte within the ovule exists in the ovary, which is an enclosed structure; in gymnosperms, the female gametophyte is there on exposed bracts of the female cone and isn't enclosed in an ovary. Double fertilization is a key occasion within the life cycle of angiosperms, however is nowhere present in gymnosperms.

In gymnosperms, the male and female gametophyte systems occur on separate male and female cones, while in angiosperms, they are part of the flower. Additionally, wind performs a vital function in pollination in gymnosperms due to the fact that pollen is blown through the wind to land at the female cones. Animal pollination is more common although many angiosperms are wind-pollinated.

6.15. SEXUAL REPRODUCTION IN ANGIOSPERMS

Angiosperms can be monoecious or dioecious and go through sexual reproduction.

The lifecycle of angiosperms undergoes the alternation of generations. The haploid gametophyte alternates with the diploid sporophyte for the duration of the sexual reproduction procedure of angiosperms. Flowers comprise the plant's reproductive structures.

6.15.1. Flower Structure

A typical flower has four primary parts, or whorls: the calyx, corolla, androecium, and gynoecium. The outermost whorl of the flower has green, leafy structures referred to as sepals, collectively called the calyx, and assist to guard the unopened bud. The second whorl is constituted of petals, generally brightly colored, collectively referred to as the corolla.

The range of sepals and petals varies relying on whether or not the plant is a monocot or dicot. Together, the calyx and corolla are called the perianth. The third whorl consists of the male reproductive systems and is called the androecium.

The androecium has stamens with anthers that include the microsporangia. The innermost group of structures within the flower is the gynoecium, or the female reproductive component(s). The carpel is the singular unit of the gynoecium and has a stigma, style, and ovary. A flower might also have one or more than one carpel.

If all four whorls are present, the flower is said to be complete. If any of the four elements is missing, the flower is known as incomplete. Flowers that include each an androecium and a gynoecium are referred to as perfect, androgynous, or hermaphrodites. Two kinds of incomplete flora exist: staminate flowers include just an androecium; and carpellate flora have just a gynoecium. If the same plant (e.g., corn or peas) bornes both male and female flowers, the species is referred to as monoecious (meaning "one home"). Species with male and female flowers borne on separate plants (e.g., *C. papaya* or Cannabis)are referred to as dioecious, or "two homes." The ovary, which may also include one or a couple of ovules, can be located above different flower elements (called superior); or it could be located under the alternative flower elements (called inferior).

6.15.2. Male Gametophyte

The male gametophyte develops and achieves maturity in an immature anther. Development of pollen in a plant's male reproductive organs occurs in a structure referred to as the microsporangium. The microsporangia, generally bi-lobed, are pollen sacs wherein the microspores grow to be pollen grains.

The microspore mother cell in the microsporangium divides through meiosis to produce four microspores, each of which will develop into a pollen grain. An internal layer of cells, referred to as the tapetum, offers nutrition to the growing microspores, contributing key components to the pollen wall.

Mature pollen grains comprise two cells: a generative cell and a pollen tube cell. The generative cell is contained in the large pollen tube cell. The tube cell develops the pollen tube upon germination via which the generative cell migrates to go into the ovary. During its transit in the pollen tube, the generative cell divides to give rise to male gametes. Upon maturity, the microsporangia burst, releasing the pollen grains from the anther.

Each pollen grain has two coverings: the exine (thicker, outer layer) and the intine. The exine consists of sporopollenin, a complex waterproofing substance provided through the tapetal cells. Sporopollenin permits the pollen to survive in adverse situations and to be carried by wind, water, or organic marketers without undergoing damage.

6.15.3. Female Gametophyte (Embryo Sac)

The standard development of the female gametophyte has two separate phases. First, in the process of megasporogenesis, a single cell within the diploid megasporangium undergoes meiosis to create four megaspores, among which only one survives.

During the second phase, mega gametogenesis, the surviving haploid megaspore undergoes mitosis to produce an eight-nucleate, seven-cell female gametophyte, also referred to as the megagametophyte, or embryo sac. The polar nuclei move to the equator and fuse, forming a single, diploid central cell. This central cell later fuses with a sperm to give rise to the triploid endosperm. Three nuclei position themselves at the end of the embryo sac opposite the micropyle and grow to be the antipodal cells, which later degenerate. The nucleus closest to the micropyle turns into the female gamete, or egg cell, and the two adjoining nuclei turn into synergid cells.

The synergids assist the pollen tube for a successful fertilization, and then they disintegrate. Once fertilization is complete, the ensuing diploid zygote develops into the embryo; the fertilized ovule forms the other tissues of the seed.

The megasporangium and, later, the embryo sac is protected by a double-layered integument. The integument will turn into the seed coat after fertilization, protecting the entire seed. The ovule wall becomes a part of the fruit. The integuments, even as protecting the megasporangium, now no longer enclose it completely, but leave a gap known as the micropyle. The micropyle permits the pollen tube to go into the female gametophyte for fertilization.

6.16. CONCLUSION

This chapter revolves around the organs and their function in water plants, the plant body, plant tissues and plant organ systems. It illustrates the meristems tissue and plant development, meristem zones. It deals with genetic control of flowers, its development, along with explaining the vegetative parts of the plant such as the roots and structure.

It also puts emphasis on the types of roots, how the roots grow, function of roots as food. The stem terminology, structures that is vascular system and the nodes, types of stems such as specialized aboveground stems and specialized below ground stems.

It gives a clear picture of the movement of water and minerals in the xylem, control of transpiration, nutritional requirements of plants and the transportation of photosynthesis in phloem. It also illustrates the chemical composition of plants that is water, nutrients, essential nutrients for plants etc.

Towards the end, it also intends to explain the plant reproductive development and structure, sexual reproduction in gymnosperms and in angiosperms, also it depicts the contrast between angiosperms and gymnosperms.

REFERENCES

1. Haynes, R., (1988). Reproductive Biology of Selected Aquatic Plants. *Annals of the Missouri Botanical Garden,* [online] *75*(3), 805. Available at: https://www.jstor.org/stable/2399368?seq=1 (accessed on 31 July 2021).
2. Lumen Learning, (n.d.). Nutritional Requirements of Plants | *Boundless Biology*. [online] Courses.lumenlearning.com. Available at: https://courses.lumenlearning.com/boundless-biology/chapter/nutritional-requirements-of-plants/ (accessed on 31 July 2021).
3. Lumen Learning, (n.d.). Plant Development | *Boundless Biology*. [online] Courses.lumenlearning.com. Available at: https://courses.lumenlearning.com/boundless-biology/chapter/plant-development/ (accessed on 31 July 2021).
4. Lumen Learning, (n.d.). Plant Reproductive Development and Structure | *Boundless Biology*. [online] Courses.lumenlearning.com. Available at: https://courses.lumenlearning.com/boundless-biology/chapter/plant-reproductive-development-and-structure/ (accessed on 31 July 2021).
5. Lumen Learning, (n.d.). The Plant Body | *Boundless Biology*. [online] Courses.lumenlearning.com. Available at: https://courses.lumenlearning.com/boundless-biology/chapter/the-plant-body/ (accessed on 31 July 2021).
6. Lumen Learning, (n.d.). Transport of Water and Solutes in Plants | *Boundless Biology*. [online] Courses.lumenlearning.com. Available at: https://courses.lumenlearning.com/boundless-biology/chapter/transport-of-water-and-solutes-in-plants/ (accessed on 31 July 2021).
7. Oregon State University, (n.d.). *Vegetative Plant Parts*. [online] OSU Extension Service. Available at: https://extension.oregonstate.edu/gardening/techniques/vegetative-plant-parts (accessed on 31 July 2021).

Chapter 7

Aquatic Plants for Water Treatment and Resource Recovery

CONTENTS

This chapter throws light on how the aquatic plants is used for the water treatment and resource recovery, it begins with explaining the wastewater treatment systems, and its various ways. It tells about the utilization of aquatic plants to treat wastewater, it defines about the artificial wetlands, potential use of aquatic plants for treating wastewater.

It also illustrates the wastewater treatment done with the help of floating aquatic plants then goes on describing the efficiency of five selected aquatic plants. It also describes the cultivation of aquatic plants for biofiltration then goes on explaining the uses of small sewage treatment plants, the designs made and the water resource recovery facilities.

7.1. INTRODUCTION

Expanding urbanization, industrialization and over populace are the variables principally liable for adding dangerous components in water, which essentially comprises substantial metals and chemicals and so forth. Water bodies are the primary targets for disposing the pollutants directly or indirectly.

This is an audit paper delineating the job of plants to help the treatment of industrial and residential wastewater. The overall purification technologies used to eliminate the foreign substances are too exorbitant and at times non-eco-friendly too.

In this manner, the research is arranged towards minimal expense and eco-friendly innovation for wastewater purification, which will be gainful for the local area. The paper examines the capability of various interaction and usage of earthbound and oceanic plants in filtering water and wastewater from various sources.

Wastewater is the mix of fluid or water-conveyed waste starting in the sterile comforts of residences, business or modern offices and organizations, notwithstanding any groundwater, surface water and tempest water that might be available.

Untreated wastewater, for the most part, contains significant degrees of organic material, various pathogenic microorganisms, just as supplements and poisonous mixtures. It hence involves environmental and health hazards and, therefore, should promptly be passed on away from its age sources and treated suitably before conclusive disposal.

A definitive objective of wastewater management is the assurance of the climate in a way equivalent to general wellbeing and financial concerns. Wastewater treatment is getting considerably more basic because of diminishing water resources, expanding wastewater removal costs and stricter release guidelines that have brought down allowable toxin levels in waste streams. The municipal area burns through huge volumes of water, and therefore produces significant measures of wastewater release.

7.2. WASTEWATER TREATMENT SYSTEMS

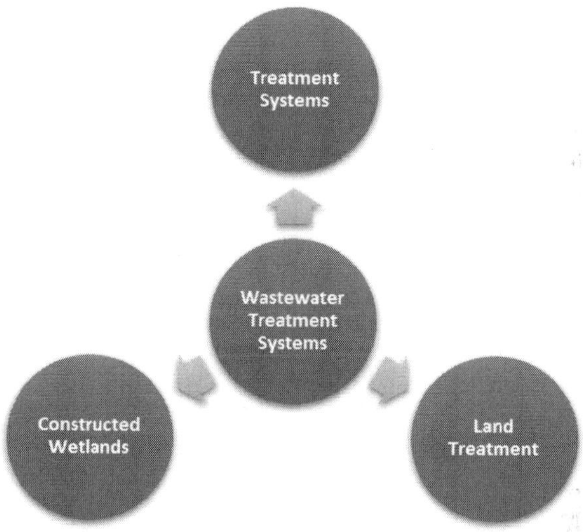

Figure 7.1: Wastewater treatment systems.

7.2.1. Treatment Systems

Regular frameworks for wastewater treatment are intended to exploit the physical, chemical, and organic cycles that happen in the common habitat when water, soil, plants, microorganisms, and the climate interact normal treatment frameworks incorporate land treatment, floating oceanic plants and built wetlands.

All regular treatment frameworks are gone before by some type of mechanical pretreatment for the expulsion of gross solids. Where adequate land reasonable for the object is accessible, these frameworks can frequently be the most practical choice as far as both development and activity.

7.2.2. Land Treatment

Land treatment is the controlled application of wastewater to the land at rates viable with the characteristic physical, synthetic and organic cycles that happen on and in the dirt. The three fundamental sorts of land treatment frameworks utilized are slow rate (SR), overflow (OF), and rapid infiltration (RI) frameworks.

7.2.3. Constructed Wetlands

Wetlands are immersed land territories with water depths regularly under 2 ft (0.6 m) that facilitates the development of rising plants like cattail, bulrush, reeds, and sedges. The vegetation gives surfaces to the connection of microbes films, helps in the filtration and adsorption of waste-water constituents, moves oxygen into the water segment and controls the development of algae by confining the infiltration of sunlight.

Figure 7.2: An illustration of cattail plant.

Source: Image by Public Domain Pictures.

Wetlands additionally clean the water by sifting through sedimentation, breaking down vegetative matter and changing over synthetic substances into the useable structures. The capacity of wetlands to reuse supplements makes them basic in the general working of earth.

7.2.4. Floating Aquatic Plants

This framework is like the FWS framework aside from the fact that the plants utilized are of the floating type, like hyacinths and duckweeds. Water depths are more prominent than on account of wetland frameworks, going from 1.6 to 6.0 feet (0.5–1.8 m). The floating plants safeguard the water from daylight and decrease the development of algae.

Figure 7.3: An illustration of floating aquatic plants.

Source: Image by Public Domain Images.

Frameworks of this sort have been compelling in diminishing BOD, nitrogen, metals and follow organics and in eliminating algae from tidal ponds and adjustment lake effluents. Supplementary aeration has been used with floating plant frameworks to expand treatment limits and to keep up the aerobic conditions essential for the natural control of mosquitoes.

7.3. UTILIZATION OF AQUATIC PLANTS TO TREAT WASTEWATER

A few papers alluded to the limit of water plants to extricate plant supplements from the water in which they developed. The explanation which has incited the spate of papers inside the most recent ten years on the supplement extraction prospects of oceanic plants is the expanding attention to the issues of water contamination – both fresh and saltwater – as an outcome of populace development and industrial development, and the disposal of human, animal and industrial waste into inland waters and into the ocean.

Numerous instances of the overwhelming results of the wastewater on in the formerly spotless and helpful waterways and lakes have excited public and scientific awareness with the need not only to exclusively capture the act of direct unloading but also to attempt to reverse it by extracting the poisons. The noteworthy capacity of an aquatic plants, especially the water hyacinth to remove mixtures and components from water effectively has gotten popular.

It is one of the earliest to go into the subject of lessening the contamination of lakes by the collecting of aquatic plants, which have removed nutrients from the water. It is called attention to that all sea-going plants can fill this need; however, little plants, similar to phytoplankton, or lowered plants are more troublesome and costly to reap than the floating and rising vascular plants.

Four species are considered appropriate: *Eichhornia crassipes* (water hyacinth), *Alternanthera philoxeroides*, *Justicia Yankee folklore,* and *Typha latifolia*. It is expressed that water hyacinth would be undeniably appropriate for nutrient removal frameworks. As it floats on a superficial level and isn't rooted, collecting is worked with.

There would be extensive microbial movement underneath the hyacinths and nutrients would be consumed by these living beings. Likewise, extensive natural matter would arrive at the water by the deficiency of root parts and presumably have a genuinely high biological oxygen demand (BOD), and it may prove to be important to utilize regular sewage holding lakes to diminish the BOD before conclusive delivery.

Alternanthera philoxeroides could presumably be best reaped by depleting the lake and afterward utilizing changed forage harvesting equipment. Typha latifolia as a rising species could be filled in lakes around 1 m deep so the water stays anaerobic, permitting base soils to eliminate P from solution.

Yet, it would be better, if space allowed, to develop the plants in water 15–20 cm deep to expand soil retention of P. At a phase when P equilibrium had been arrived at then the lakes could be dried and utilized for traditional yields until the P levels are diminished.

Figure 7.4: An illustration of alternanthera philoxeroides flower head.

Source: Image by Flickr.

Nutrient removal by plants would be best when the nutrient is concentrated. This is typical of the effluents from feed heaps of cattle. In the event that plants were developed on the lakes holding these effluents, they could be fed to the cattle. This would be ideal since taking care of and transport of the plants would be kept to a minimum. Nutrient removal and plant usage could likewise be dealt with by a similar association.

A strategy for gathering aquatic weeds stresses the value of oceanic plants in filtering water. He remarks that water plants are perhaps the most effective methods of removing nutrients from water. By releasing oxygen, they additionally add to water lucidity.

In the field and lab assessments of bioassays for nitrogen and phosphorus with algae and oceanic weed, method depicts the states of excess, or restricting concentrations of nitrogen and phosphorus in algae and aquatic plants.

The report likewise talks about a straightforward technique for estimating paces of nitrogen concentration by blue-green algae. The test plants utilized were *Ceratophyllum* sps., *Cladophora* sps., *Hydrodictyon* sps., *Aphanizomenon* sps., *Microcystis* sps., *Anabaena* sps., and *Nostoc* sps.

It is tracked down those aquatic plants which have been as of late presented to a generally high convergence of accessible P could assimilate enough to show overflow P conditions by their absolute P content.

A similar examination on nutrient substance of aquatic plants from various living spaces was finished. In the trial, three oceanic plants were developed in water with various levels of contamination, from crude sewage to unpolluted.

The take-up of supplements was dissected. "Luxury consumption" of nutrients was seen from water containing a lot of N, P and K. The creators close, "As the accessibility of the nutrient changes so does the concentration of nutrient in the plant."

The relationship of nutrients in water to algae and aquatic plants was noticed. As needs be, "The development and advancement of established aquatic plants will rely upon the accessible nutrients in the water, appropriate climatic conditions and rivalry with different species.

The void made by the obliteration of one kind of organic development might be filled by another and distinctive sort of natural action." The creator audits the relationship and significance of aquatic plants to their natural surroundings. A few oceanic plants have been found to gather arsenic to a level well over those related with poisonousness of this component.

The impact of sewage effluent on the development of five aquatic species. *Eichhornia crassipes*, *Alternanthera philoxeroides*, *Egeria densa*, *Najas flexilis,* and *Potamogeton crispus* were filled in plastic pools in well water, with or without the expansion of 25% of sewage effluent.

Figure 7.5: An illustration of Egeria densa.

Source: Image by Wikimedia commons.

Of the five test plants, *E. crassipes* showed the most extreme development reaction to the sewage gushing, with *A. philoxeroides* second. In well water alone *A. philoxeroides* was the solitary plant to endure, showing its capacity to endure low nutrient levels.

The water hyacinth ruled others covering 71% of the water surface and eliminated 6.9 g of N, 2.9 g of P and 8.7 g of K from the sewage pools. *A. philoxeroides* didn't acquire N and K, rather the plants delivered these components to the water; however, they took up some P (0.15 mg).

The significance of wetlands has changed after some time. Wetlands fill in as change zones among aquatic and earthly conditions and give a powerful connection between the two. Moving along an inclination, water gets synthetics and sediments which, as they travel through the wetland, are changed and shipped to the encompassing scene.

The biotechnological answer for wastewater includes introducing fake wetlands to go about as regular channels. By finding built wetlands between the wastewater source and oceanic assets (streams, lakes, tidal ponds), these frameworks require no support, burn-through no power, and cost short of what one-quarter that of a customary waste treatment framework. Such built wetlands utilize various types of plants that regularly have large amounts of normal wetlands: cattails, water lilies, and rushes.

7.3.1. Artificial Wetlands for Sewage and Industrial Waste

1. Sewage streams into the constructed wetland, which is an unearthed cell loaded up with sand that serves to sift through smells.
2. The filter comprises of a huge vegetative planting, for this situation rushes, whose roots in the sand are taken care of by the wastewater.
3. The nutrients in the water are consumed by rushes (Juncus), which sequester them in their tissues as they develop.
4. The nutrients ingested are killed with vegetation dieback of the rushes, whose leftovers structure a protecting layer.
5. The purged water channels from the wetland into the tidal pond.
6. Proportioning a wastewater treatment wetland: The region required is relative to the size of the private populace and is determined as follows: 1 individual = around 5 m^2.

Treatment frameworks utilizing oceanic plants comprise of shallow supplies containing coasting or lowered aquatic plants. The best-considered wastewater frameworks are those using duckweed (Lemna minor). For the most part, treatment frameworks break into two sorts dependent on the predominant plant types.

The main sort utilizes floating plants which are recognized by their capacity to address their issue for carbon dioxide and oxygen straightforwardly from the air. Such plants get their mineral necessities from the water.

The second sort of treatment framework comprises of submerged plants, which are recognized by their capacity to assimilate oxygen, carbon dioxide, and minerals straightforwardly from the water segment. Submerged plants are effectively repressed by high turbidity on the grounds that their photosynthetic parts are submerged. Barely any example cases are referenced underneath:

Maria Alejandra Maine united a group from UNL to research strategies for treating the effluents of the Bahco Company. They built a wetland on the grounds of the plant in Sao Tome – Province of Santa Fe Argentina, to play out the last treatment of the organization's gushing of the two liquids and mechanical waste.

Building an outside wastewater treatment framework using aquatic plants to ingest and clean fluid waste emerged in 2001 when Bahco Argentina chose to improve its natural administration technique. To limit the effect produced by assembling measures, the organization chose to add a last cleaning stage to its wastewater treatment framework.

7.4. ARTIFICIAL WETLANDS

An artificial wetland is an arrangement of treating wastewater in a shallow developed lake or channel close to 0.60 meters down, in which oceanic plants have been planted, and common cycles are used to treat wastewater. Fake or developed wetlands enjoy the upper hand over elective treatment frameworks in that they require almost no or very little amount of energy to operate.

In the event that there is sufficient modest land accessible close to the emanating source, establishment of the water treatment wetlands can be a savvy elective. Likewise, developed wetlands give living space to natural life, and are tastefully satisfying to the eye.

As such, a developed wetland or a fake wetland is a natural wastewater treatment framework that copies and improves the adequacy of the cycles that help to decontaminate water like normally happening wetlands. The framework utilizes water, aquatic plants (i.e., reeds, duckweed), normally happening microorganisms and a channel bed (typically of sand, soils or potentially rock).

Built wetlands can be utilized for one or the other optional or tertiary wastewater treatment. Various plans exist, including vertical wetlands, which require less land, yet more energy for tasks like siphoning or directing than level wetlands, which can rather depend on gravity and geology.

The broad alternatives in plan, materials and innovation permit the developed wetland to be adjusted to nearby conditions and land accessibility. Expenses are reliant upon the cost of land and materials, yet where land is less expensive and generally accessible, developed wetlands are an extremely financially savvy technique for wastewater treatment.

The overall idea is that the plants, microorganisms and substrates together go about as a channel and sanitization framework. In the first place, water is eased back as it enters the wetland, taking into account the sedimentation of solids.

Through the cycle of watercourse through the developed wetland, plant roots and the substrate eliminate the bigger particles present in the wastewater. Contaminations and nutrients present in the wastewater are then normally split down and taken up by the microbes and plants, accordingly eliminating them from the water.

The maintenance time in the wetland, which changes relying upon the plan and wanted quality level, alongside UV radiation and plant emission of anti-toxins, will likewise kill the microorganisms present in wastewater. After treatment in a built wetland, water can be securely delivered into surface waters or utilized for different purposes.

It has some of the salient features:

1. Effectively treats wastewater from human waste, horticultural spillover, stormwater, and a few metals or contaminations from mining and industry;
2. Uses innovation that is easy to comprehend and oversee;
3. Cost productive as far as development, tasks, and upkeep;
4. Low energy utilization needed for activities;
5. Prepares water for reuse;

6. Assists in keeping up groundwater and surface water levels;
7. Contributes to natural assurance by giving an environment to plants and creatures
8. Acts as a method for water stockpiling;
9. Pleasing common feel.

A portion of its benefits are:

1 .Plants can be utilized as minimal expense extraction gadgets to purge dirtied water.
2. In a few cases, plants disintegrate squander quicker than microorganisms.
3. The strategy can be applied to enormous territories or to finish the disinfecting of limited regions in protracted periods.

Disadvantages are:

1. The cycle is restricted to shallow water or the profundity to which roots can infiltrate.
2. The interaction can be tedious.
3. There is a biological limit to the metals and mixtures that can be caught.

7.4.1. Role of Aquatic Plants in the Treatment of Waste Water by Artificial Wetlands

The limit with respect to wastewater cleansing by both common and artificial wetlands is all around reported. Wetlands eliminate aquatic contaminations through an intricate assortment of natural, physical and chemical measures.

Albeit the higher aquatic plants are the clearest biological component of the wetland biological system, ongoing reports in the writing, point to the way that uptake of toxins by wetland vegetation can't, without help from anyone else, represent the high contamination expulsion efficiencies frequently saw at the high stacking rates normal for some treatment circumstances.

Maybe, the significant components for toxin expulsion in these wetland frameworks incorporate both bacterial changes and physico-compound handling including adsorption, precipitation and sedimentation. This isn't to say in any case, that the higher plants don't assume a significant part in wastewater sanitization.

Specifically, the plant rhizome gives surfaces to bacterial development just as for filtration of solids. All the more significantly, plants are known to move oxygen (O_2) from the shoots to the roots. The rhizosphere or rootzone will in this way offer an oxidized microenvironment in a generally anaerobic substrate, which animates both the disintegration of natural matter and the development of nitrifying microscopic organisms, the last which can change alkali over to nitrate.

The nitrate so shaped would then be able to diffuse or permeate through to the oxygen-helpless zones where it will be eliminated from the framework by denitrification.

While a couple of studies have been accounted for on the connection between the oxidizing capacity of oceanic plant roots and the microbial cycles (counting nitrogen changes) happening in the substrate, there stays an absence of data on the quantitative job of an assortment of higher plants in real treatment circumstances where either common or fake wetlands are being utilized for wastewater cleaning.

7.5. POTENTIAL USE OF AQUATIC PLANTS FOR TREATING WASTE WATER

The possible employments of developing aquatic plants for wastewater refinement were analyzed on a laboratory scale in group tests. These aquatic plants, duckweed, water hyacinth and green growth (*Chlorella vulgaris*), improves the evacuation of contaminations by retaining as plant nutrients.

Plants are fit for diminishing all tested indicators of water quality to even out that license the utilization of cleansed water for irrigation system. Compound oxygen interest (COD) and organic oxygen interest (BOD) decreased in wastewater to (43% and 42%) by duckweeds, (28% and 33%) by water hyacinth, and (33% and 38%) by green growth for 21 days.

Duckweeds have a lot higher poisons expulsion efficiencies, particularly, N, P, K, and other hefty metals than water hyacinth and green growth. Thus, duckweed was guarantee to use in wastewater medicines than water hyacinth and green growth.

Crude sewage contains complex natural and inorganic materials, including proteins, urea, amines, cellulose, fats, sugars and cleanser. In natural treatment offices, microbes, organisms, zooplankton, and green growth debase and utilize these mind-boggling materials, bringing about an emanating wealthy in nitrogen, potassium, phosphorus, and different components.

The treated waste has ordinarily been released into the closest lake or stream with little thought about the outcome, today, nonetheless, the developing total populace and expanding interest for an improved climate make better disposal mandatory.

Some oceanic weeds, for example, duckweed and water hyacinth, were found to have the capacity of refining the natural toxins and taking up inorganic nutrient salts, for example, nitrogen and phosphorus compounds.

Exploration in the course of recent many years has demonstrated that some floating plants, for example, water hyacinth (*E. crassipes*), water lettuce (*Pistia stratiotes*), pennywort (*Hydrocotyle umbellata*), duckweed (*Lemna minor*), water nut (*Alternanthera philoxeroides*) and lidded (*Cleistocalyx operculatus*), have the best consequences for cleaning eutrophic water.

Figure 7.6: An illustration of water lettuce.

Source: Image by Wikipedia.

In this way, the macrophyte lake can be utilized as wastewater treatment offices with so many merits as low capital and working/upkeep costs, energy-saving, wide range contaminations evacuation and the recuperation of the oceanic plants as staples for creatures or readiness of green compost and the reuse of wastewater as assets.

In this examination, a preliminary research center was made to evaluate the toxin expulsion from wastewater in the macrophyte lake of El-Nazlah wastewater treatment plant at El-Fayum Governorate, Egypt utilizing duckweeds, water hyacinth and green growth as guide for use of macrophyte lake.

7.6. WASTEWATER TREATMENT THROUGH FLOATING AQUATIC PLANT SYSTEMS

Aquatic plant systems are designed and developed systems that utilize aquatic plants in the treatment of modern or homegrown wastewater. They are intended to accomplish a particular wastewater treatment objective. Aquatic plant systems can be separated into two classifications:

1. Systems with floating aquatic plants, for example, water hyacinth, duckweed, and pennywort.
2. Systems with submerged aquatic plants like waterweed, watermilfoil, and watercress.

The utilization of hydroponics as a method for treating wastewater includes both characteristic and counterfeit wetlands and the creation of green growth and higher plants (submerged and drenched) to eliminate impurities, for example, nitrogen compounds, BOD5, hydrocarbons, and substantial metals from the wastewater.

Skimming aquatic plants, for example, water hyacinth (*Eichhornia crassipes*) and duckweed (*Lemna* spp.) seem, by all accounts, to be probably the most encouraging aquatic plants for the treatment of wastewater and have gotten the most consideration in such manner. Different plants are likewise being considered, among them ocean growth and gator weed.

Figure 7.7: An illustration of Eichhornia.

Source: Image by Flickr.

These systems are essentially shallow lakes covered with floating plants that confine wastewater for at least one week. The fundamental reason for the plants in these systems is to give appropriate environments to microscopic organisms that eliminate by far most of dissolved nutrients.

These advancements are valuable in regions where appropriate plants are promptly accessible. In territories where they are not, any presentation of plant species should be embraced with caution to limit the chance of making disturbance in development conditions.

In any event, bringing them into built enclosures ought to be done cautiously, and with the premonition that there is a solid probability that they will enter characteristic water systems (particularly on the grounds that they should be reaped from the treatment systems and discarded).

The water hyacinth is an enduring, free-coasting freshwater aquatic macrophyte with adjusted, upstanding, thick, waxy, and reflexive green leaves and spikes of lavender blossoms, local to South America and discovered normally in streams, narrows, and different backwaters in mild and tropical areas.

The water hyacinth is viewed as one of the most exceedingly awful weeds on the planet—aquatic or earthly—for its quick development, which will in general stop up the streams for boat traffic and keep daylight and oxygen from getting into the water.

It flourishes in nitrogen-rich conditions and therefore does incredibly well in crude and mostly treated wastewaters. At the point when it is utilized for profluent treatment, wastewater is gone through a water-hyacinth-covered bowl, where the plants eliminate nutrients, suspended solids, hefty metals, and different foreign substances.

Group treatment and move through systems, utilizing single and different tidal ponds, are utilized. In light of its quick development rate and natural protection from creepy crawly predation and illness, water hyacinth plants should be gathered from these systems. Albeit numerous employments of the plant material have been researched, it is for the most part suggested as a wellspring of methane when anaerobically processed.

Its utilization as a manure or soil conditioner (subsequent to fertilizing the soil), or as a creature feed, is frequently not prescribed attributable to its penchant to collect weighty metals. The plant likewise has a low natural substance (it is fundamentally water) and, when treated the soil, leaves behind minimal material with which to enhance the dirt.

Duckweed (*Lemna* sp., *Spirodela* sp., and *Wolffia* sp.) are little, green freshwater plants with fronds from one to a couple of millimeters and a short root, normally under 1 cm long. Duckweed are the littlest blossoming plants. They develop as little states of plants coasting on the surfaces of calm waterways. Their development can be incredibly fast, given the legitimate conditions.

These plants are practically all leaf, having basically no stem tissue, and just one, or a couple, fine roots. In nature, duckweed fills in as nourishment for some types of fish and aquatic birds. In the United States, crawfish are frequently delivered in flooded rice fields in rice-development spaces of the United States to control weeds (regularly duckweed).

Since they are high in proteins and furthermore (3-carotene, reaped duckweed can be utilized as feed for grass carp, sheep, chicken, and tilapia. Duckweed can endure and develop under a wide scope of conditions, remembering for water dirtied with high groupings of microbes and some horticultural squanders. These qualities have carried duckweed to the consideration of ecological architects and aquaculturists the same.

Water hyacinth systems are mostly floating aquatic plant systems for wastewater treatment. Three kinds of the systems exist today for BOD5 and nutrient evacuations: high-impact non circulated air through, vigorous circulated air through, and facultative anaerobic.

These kinds of water hyacinth systems mirror the sorts of adjustment lakes utilized for wastewater treatment. As anyone might expect, floating aquatic plants, for example, water hyacinth, for wastewater treatment flourish in regular or counterfeit lakes and built wetlands.

Notwithstanding, utilization of water hyacinth has been restricted, in geographic area, to warm climate areas due to the affectability of water hyacinth to freezing conditions. Water hyacinth systems have been frequently utilized for either eliminating green growth from oxidation lake effluents (by obstructing the daylight to arrive at the water) or for nutrient expulsion following auxiliary treatment.

Plant models for wastewater treatment utilizing water hyacinth incorporate the profundity of the tidal ponds, which ought to be adequate to expand root development and the ingestion of nutrients and hefty metals; detainment time; the stream rate and volume of emanating to be dealt with; and the ideal water quality and likely employments of the treated water.

Land prerequisites for lake development are roughly 1 m^2/m^3/day of water to be dealt with. Phosphorus decreases acquired in such systems range somewhere in the range of 10% and 75%, and nitrogen decreases somewhere in the range of 40% and 75% of the influent focus.

Duckweed systems have been utilized effectively to improve the emanating quality from facultative adjustment lakes by diminishing green growth populace, BOD5, and nutrients. Duckweed is touchy to wind and might be blown in floats to the leeward side of the lake, bringing about openness of a huge surface territory that is inclined to green growth blossoming.

Rearrangement of the plants to cover the surface requires difficult work. Heaps of disintegrating plants can bring about the creation of smells. Float control, like coasting astounds, can be utilized to isolate the surface territory into more modest cells, accordingly lessening the measure of open surface region to blowing wind.

Like water hyacinth systems, design rules of duckweed systems incorporate pressure driven detainment time, water profundity, lake math, BOD5 stacking, and pressure driven stacking rate. Normal treatment results from a few U.S. areas, including duckweed wastewater treatment systems for effluents from facultative adjustment lakes are given.

7.7. EFFICIENCY OF FIVE SELECTED AQUATIC PLANTS IN PHYTOREMEDIATION OF AQUACULTURE WASTEWATER

The absence of clean water sources, because of the presence of toxins in water, is a significant issue in numerous nations, including Malaysia. To conquer this issue, different strategies have been presented, including phytoremediation treatment.

Consequently, this phytoremediation study inspected the capacity of five aquatic plants—*Centella asiatica, Ipomoea aquatica, Salvinia molesta, Eichhornia crassipes*, and *Pistia stratiotes*—to eliminate three toxins—absolute suspended solids (TSS), ammoniacal nitrogen (NH3-N), and phosphate—from hydroponics wastewater.

Utilizing wastewater tests, each containing 50 g of one of the plants, the poison levels were estimated at regular intervals for 14 days. The outcomes showed an exceptional decrease in the convergence of poisons, where *C. asiatica* had the option to eliminate 98% of NH3-N, 90% of TSS, and 64% of phosphate, while *I. aquatica* showed the possibility to wipe out up to 73% of TSS and NH3-N, and half of phosphate.

E. crassipes radically eliminated 98% of phosphate, 96% of TSS, and 74% of NH3-N, while *P. stratiotes* had the option to kill 98% of TSS, 78% of NH3-N, and 89% of phosphate. *S. molesta* was proficient in eliminating 89.3% of TSS and 88.6% of phosphate, yet just eliminated 63.9% of NH3-N

Aquaculture is the breeding, rearing, and harvesting of aquatic life, like fish, shellfish, prawns, green growth, and different life forms living in an aquatic biological system. Aquaculture is one of the quickest developing economies on the planet food economy.

Worldwide aquaculture creation has multiplied in the previous decade as far as worth and volume, and now supplies 33% of the fish devoured overall. Asia rules this creation, representing 88.91% by volume in 2014. There is no uncertainty that the commitment of aquaculture to the food supply will increment later on because of the developing total populace.

Be that as it may, unreasonable aquaculture can lead to new issues, for example, water contamination, if the emanating and wastewater are not fastidiously kept up. Water contamination brought about by gushing from lakes is one of the principal issues identified with aquaculture. Aquaculture has gotten huge enough to essentially affect the climate, and some ecological concerns have been communicated by natural activists and researchers.

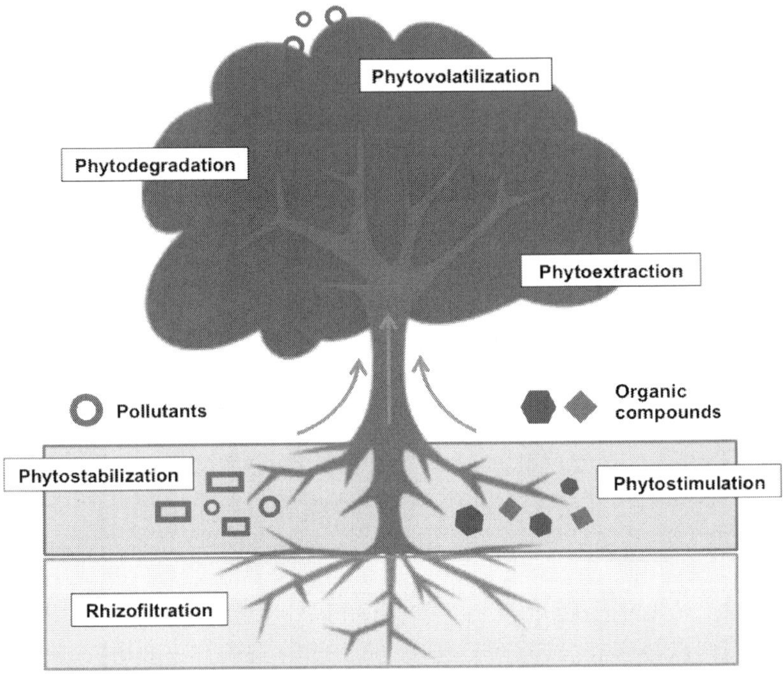

Figure 7.8: An illustration of phytoremediation process.

Source: Image by Wikimedia Commons.

In most freshwater aquaculture lakes, wastewater is produced prior to gathering the stock, in which the entirety of the particles and supplements are available. Potential toxins come from composts, food spills, and the misuse of raised creatures. A portion of the significant wellsprings of wastewater contamination are leftover food squander, dung, metabolic side-effects, and remaining biocides and biostats.

These items contain natural carbon, nitrogen, alkali, urea, bicarbonate, phosphate, nutrients, and therapeutics, which, in high fixations, are hurtful to the aquatic environment and in the long run upset the common evolved way of life. As per Hanafiah et al., wastewater containing these substances will have significant adverse consequences on human wellbeing just as the climate. This will make significant unsettling influence our biological system balance in the end, making the Millennium Development Goals (MDGs) difficult to accomplish.

There is a great deal of regular techniques that assistance to eliminate contaminations from the aquatic climate like adsorption, natural therapies, sterilization, bright light, and progressed treatment advances like progressed oxidation measures (AOPs).

Albeit these strategies can altogether improve the evacuation of poisons, the capital, operational, and support costs for these medicines are high, particularly for cutting edge medicines. Phytoremediation, in any case, is one of the harmless to the ecosystem and minimal expense techniques for eliminating toxins from water and soil.

This treatment is a remediation strategy that utilizes different plants to decrease, concentrate, and hold exorbitant supplements and toxins from a polluted medium. Phytoremediation strategies can be sorted dependent on how the poisons are taken out from the climate, for example, through corruption, extraction, regulation or a mix of at least two methods.

Plant species viewed as ideal phytoremediation specialists have a high biomass, a decent root system, and are quickly developing and simple to develop.

7.8. CULTIVATION OF AQUATIC PLANTS FOR BIOFILTRATION OF WASTEWATER

Effective wastewater management has a fundamental part in observing the flood for danger relief and furthermore people groups' prosperity through the prevention of infection disease. Wastewater is by and large created through the family exercises, cultivating works, and assembling measure.

Wastewater should be treated before release into the normal water sources or contaminate the water supply. In this examination study, aquatic plants were utilized for the biofiltration of treated homegrown wastewater before release into the climate.

The outcomes acquired from the investigation exhibited that the pH has no critical change (or even possibly). The pH of the wastewater test was named protected before the treatment. In any case, the decrease of turbidity, nitrate, phosphorous, and smelling salts utilizing the biofiltration procedures yielded positive outcomes. Consequently, biofiltration is a brilliant option in contrast to tertiary wastewater treatment.

Productive wastewater management has an indispensable job in observing the flood for risk relief and the prosperity of individuals. Wastewater is for the most part created through family exercises, cultivating works, and assembling measures. Wastewater should be treated before release to the characteristic water sources, or it will contaminate the water supply.

Wastewater should be passed on to the wastewater treatment plant to be dealt with, and it must be passed on through the seepage frameworks. Waste frameworks comprise of a mind-boggling foundation like lines, siphons, screens, and channels. A few nations on the planet, principally Latin American nations, contingent upon water trades, which as of now is increasing around 3.5 occasions at the pace of fares.

These make Mexico as the main supporter of water trades in Latin America. The interest is progressively high and is projected to proceed in the coming years, which can prompt water deficiencies whenever left unmitigated. Moreover, the utilization of macrophytes in the biofiltration cycle of wastewater is more viable in regions with warm environments.

In any case, it has all the earmarks of being an ideal alternative for created countries in light of the advantages of this cleansing interaction. Various examinations have shown that macrophytes have the ability to ingest overabundance supplements from wastewater. Pavithra and Kousar assessed the possibilities of goliath Salvinia in the treatment of material wastewater.

The result of the examination showed that goliath Salvinia is compelling in the treatment of material wastewater. Hanafiah et al., researched the possibilities of phytoremediation utilizing *P. stratiotes* and *S. molesta* in the treatment of wastewater.

The investigation showed that *P. stratiotes* and *S. molesta* are proficient in the evacuation of absolute suspended solids and ammoniacal nitrogen. Ariffin et al. (2019), examined the proficiency of *Azolla pinnata* for phytoremediation of wastewater aquaculture. The outcome acquired demonstrated that *Azolla pinnata* definitely diminished the smelling salts and phosphate content in the aquaculture wastewater.

In this investigation, three notable plants that are generally known to be utilized for biofiltration of water treatment were utilized. The plants that we utilized are Water Lettuce, Water Hyacinth, and Giant Salvinia. The plants were privately sourced and introduced into the acrylic tank.

All the chosen vegetation utilized are notable for its synthetic treatment properties and silt sifting. Among the most well-known water treatment vegetation is water lettuce, or *Pistia stratiotes*. Water Lettuce is accounted for to yield a decent outcome in separating and treating city wastewater.

Furthermore, the vegetation is liable for being a hyperaccumulator of follow metals like Cr, Cu, Fe, Mn, Ni, Pb, and furthermore Zn, alluding to the vegetation bioconcentration factor record. *Pistia stratiotes* is a typical plant that can be found worldwide because of its environment resilience, and it normally develops well inside a temperature scope of 15–35°C.

The foundations of water hyacinth (WH) normally ingest poisons just as some natural mixtures which are cancer-causing and have weakening in a district of multiple times that is typically found in the water climate.

WH can be supported for wastewater treatment, and it can assist with the method of water sanitization either for modern wastewater or sewer water, notwithstanding other accessible strategies. The root designs of WH give a reasonable climate to vigorous microscopic organisms to eliminate different pollutants present in wastewater.

Goliath Salvinia is an intrusive aquatic weed and free-coasting plant. In any case, because of the mild environment, Malaysia is more helpless to Salvinia invasion. Monster Salvinia changes in shading from green to gold to brown. It is an aquatic greenery with gliding and elliptical leaves, in which the leaves of youthful plants lie level on the water surface.

Conversely, the leaves of developed plants develop between 1.3 to 4.0m. A long chain development happens as the plants become together to shape mats. Goliath Salvinia is productive and, under good conditions, can twofold the size of its mats in 7 to 10 days. As the mats keep on developing, they structure thick layers of vegetation.

Likewise, the outside of the Salvinia leaf has columns of round and hollow "hairs" finished off with four branches that are joined at the tips to frame an enclosure or eggbeater shape. These hairs give a rich appearance and repulse water. In developed plants, submerged rootlike structures cover follows that can have egg-molded spore cases joined. In the United States, these spore cases are not known to contain ripe spores. This weed overwinters and reproduces vegetatively.

7.9. USES OF SMALL SEWAGE TREATMENT PLANTS

A sewage treatment plant is needed for all private and business territories. This interaction is like that of a septic tank. In any case, some extra mechanical parts in the previous demonstrate a superior decision bringing about cleaner and greater environment-friendly effluent. Sewage treatment, otherwise called wastewater treatment, is utilized to eliminate pollutants from city wastewater. Different physical, natural, and chemicals cycles are utilized to treat water.

Limited scope wastewater treatment is essentially for homegrown sewage. This is considered as the auxiliary treatment explicitly intended to eliminate natural toxins from water. Regular sources of these debasements are food waste, human waste, detergent, soap, and so forth.

7.9.1. Design of a Small Sewage Treatment Plant

Little sewage treatment plants are normally smaller and independent which don't need any outside support. They are above and beyond to treat wastewater from families, inns, and even towns.

A portable submersible pump is utilized to reuse the plant profluent and all tanks ought to have stainless steel stepping stool. Abundance of oil and grease may mess up the framework. Henceforth, vital precautions are required.

There are some rules which should be remembered for the sewage treatment plant plan:

1. Location of the plant comparable to the neighborhoods and environmental factors;
2. Final arrangement with area of emanating release;
3. Process outlines and subtleties plan computations;
4. Hydraulic profile;
5. Ventilation subtleties;
6. Light subtleties;
7. Equipment timetables alongside their lists and support manuals;
8. Uses of small sewage treatment plants.

Water is a restricted asset in the climate, and we need to reuse it. Enormous water treatments plants are introduced to eliminate contaminations so what the water is protected to be utilized once more. In this cycle, some reusable strong waste is likewise taken out which is known as sludge. It is utilized as a fuel source and furthermore for rural purposes.

A definitive thought of sewage treatment plants is to give ecologically safe sewage water. It is named as effluent. Little sewage plants can be redone dependent on the requirements of a specific territory which is for the most part impacted by the climatic state of the space, way of life of individuals around, topographical areas, and so forth.

The interaction begins with the underlying breakdown of natural solids. It is trailed by the effluent channel, and afterward the filtration interaction is completed. Further, with the organic treatment, microorganisms are taken out from the sewage.

Little sewage treatment plant is utilized to eliminate salts which as sodium, nitrates, phosphate, nitrogen, and so forth which are normally found in homegrown waste, basically kitchen. Microorganisms are likewise isolated through this interaction.

Water created through this interaction is utilized for water systems, hydrants, latrine flushing, cooling water, substantial water, and numerous such purposes.

Small scale sewage treatment plant measures homegrown/small scale wastewater and delivers clean water which can be utilized straightforwardly in the climate for different purposes generally agribusiness and cultivating. Customary overhauling of the plant is likewise needed for best use which keeps the mechanical segments is totally great working condition.

These medicinal plants are likewise purged intermittently it fundamentally permits the framework to be liberated from sludge which normally aggregates with time.

Other than utilizing water cautiously, it is additionally critical to reuse the asset if conceivable. Thus, small-scale sewage treatment plants are exceptionally fundamental for homegrown or small-scale water treatment measures.

7.10. WATER RESOURCE RECOVERY FACILITIES

Water is the most significant natural resource on the planet. Districts and army bases should oversee and steward wastewater from numerous likely sources, including sterile and family, business, institutional, producing, mechanical, development, and tempest water.

The executives and treatment can shift from little on-location treatment and release to discharge to large-scale water resource recovery facilities with collection systems and pumping stations, treatment processes, oil-water separators, reusing, energy recuperation, and reuse. The requirement for community wastewater assortment and treatment system worldwide is a developing one, starting more than 200 years prior in the US. At first, endeavors were centered around collection and removal and were driven by the need to lessen human illness.

That era was followed by an emphasis on the elimination of gross water contamination impacts, permitting local marine life forms to get back to ordinary development designs and permitting full human sporting use.

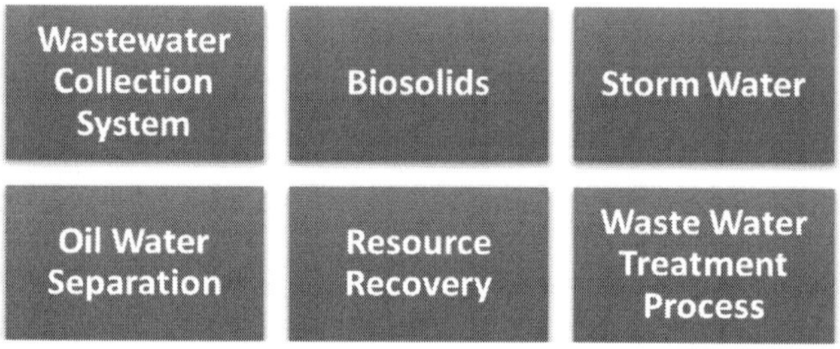

Figure 7.9: Water recovery facilities.

At long last, community wastewater collection and treatment systems have started to rethink wastewater as an asset with important items to be removed through treatment. In the United States, wastewater and tempest water treatment releases are represented by the Clean Water Act.

As populace and environment pressures increase, safe water and water system will just turn out to be more significant. As per the EPA's 2012 Clean Watersheds Needs Survey, 238.2 million individuals in the United States were adjusted by Publicly Owned Treatment Works (POTW), with an extended increment of more than 10% by 2032.

More than 300 billion cubic meters of city wastewater is created every year on the planet. The need to ensure human wellbeing, recuperate and monitor assets while adjusting to an evolving scene, including new advances and aging infrastructure, will drive further developed wastewater treatment.

With the capacity to recuperate important assets from wastewater, like phosphorus, nitrogen, and biogas, just as saving water for reuse, drinking, farming, cooling or mechanical use, POTWs are currently suitably alluded to as water asset recuperation offices as they keep on recuperating natural resources through wastewater treatment.

7.10.1. Wastewater Collection System

Assortments Systems allude to the arrangement of underground lines and support structures that transport wastewater to water asset recuperation offices. These frameworks incorporate gravity pipes, sewer vents, lift stations, power mains and more as they gather and transport private and business waste.

7.10.2. Waste Water Treatment Process

Wastewater treatment alludes to the expulsion of poisons from wastewater before release to a waterbody. This incorporates actual cycles like sedimentation and filtration, compound cycles, for example, precipitation and organic cycles like circulated air through tidal ponds or enacted slime.

7.10.3. Resource Recovery

Water asset recuperation facilities present an interesting chance to diminish energy requests, lower expenses, and increment creation of sustainable power by making important items out of customary waste streams. For water asset recuperation offices, these assets incorporate water, energy, supplements (phosphorus and nitrogen) and biosolids.

7.10.4. Biosolids

Biosolids are the supplement rich natural material coming about because of the treatment of homegrown waste at a water asset recuperation office. Through biosolids the board, strong buildup from wastewater treatment is handled to diminish or wipe out microbes and limit scents, framing a protected, helpful rural item as per administrative necessities that can be utilized as manure and different items.

7.10.5. Storm Water

Stormwater alludes to a substantial amount of water, like downpour or snow, that tumbles to the outside of the Earth which gets dirtied as it gets, conveys, and moves different toxins (oil, oil, synthetic compounds, silt, supplements, microbes) along roads, channels, open channels, and Storm sewer frameworks. A large portion of the untreated spillover at last is released into close by water bodies. Notwithstanding, in joined sewer frameworks, stormwater streams with wastewater and is treated at a water asset recuperation office.

7.10.6. Oil Water Separation

Oil-water separation alludes to the actual detachment of oil-water blends into their different segments. Oil-water separators fluctuate broadly and are picked dependent on execution boundaries and activity conditions. Legitimate oil-water division is important to forestall pollution to storm sewer and clean sewer frameworks.

7.11. CONCLUSION

This chapter gives an insight to the reader how aquatic plants are used for water treatment and resource recovery, it starts with the explanation of wastewater treatment systems, the land treatment, and constructed wetlands. It explains the functions of floating aquatic plants and how they are used for the treatment of wastewater, the artificial wetlands for sewage and industrial waste.

It elaborates the functions of artificial wetland and defines the role of aquatic plants in the treatment of wastewater by artificial wetlands along with explaining the potential use of aquatic plants for treating the wastewater then goes on explaining wastewater treatment through floating aquatic plants.

It also puts emphasis on the efficiency of five selected aquatic plants, cultivation of aquatic plants for biofiltration, uses of small sewage treatment plants, its designs. Then the water resource recovery facilities which includes wastewater collection system, treatment process, resource recovery, biosolids, stormwater, and the oil-water separation.

REFERENCES

1. Archana Dixit, Savita Dixit, & Goswami, C. S. (2011). Process and plants for wastewater remediation: a review. [ebook] *Scientific Reviews and Chemical Communications,* pp. 71–77. Available at: https://www.tsijournals.com/articles/process-and-plants-for-wastewater-remediation--a-review.pdf (accessed on 31 July 2021).
2. Centre for Science and Environment, (2021). Constructed Wetlands – Wastewater Treatment Systems. [online] Cseindia.org. Available at: https://www.cseindia.org/constructed-wetlands-wastewater-treatment-systems-6215 (accessed on 31 July 2021).
3. Cristian Frers, (2014). The Use of Aquatic Plants to Treat Waste Water | Fourth Corner Nurseries. [online] *Fourth Corner Nurseries.* Available at: https://fourthcornernurseries.com/the-use-of-aquatic-plants-to-treat-waste-water/#:~:text=wastewater%20treatment%20with%20aquatic%20plants&text=Generally%2C%20treatment%20systems%20break%20into,oxygen%20directly%20from%20the%20atmosphere. (accessed on 31 July 2021).
4. El-Din, S., & Abdel-Aziz, R., (2018). *Potential Uses of Aquatic Plants for Wastewater Treatment.* [online] Pulsus.com. Available at: https://www.pulsus.com/scholarly-articles/potential-uses-of-aquatic-plants-for-wastewater-treatment-4608.html (accessed on 31 July 2021).
5. Gersberg, R., Elkins, B., Lyon, S., & Goldman, C., (1986). Role of aquatic plants in wastewater treatment by artificial wetlands. *Water Research,* [online] *20*(3), 363–368. Available at: https://www.sciencedirect.com/science/article/abs/pii/0043135486900850 (accessed on 31 July 2021).
6. Gersberg, R., Elkins, B., Lyon, S., & Goldman, C., (1986). Role of aquatic plants in wastewater treatment by artificial wetlands. *Water Research,* [online] *20*(3), 363–368. Available at: https://www.researchgate.net/publication/223797169_Role_of_Aquatic_Plants_in_Wastewater_Treatment_by_Artificial_Wetlands (accessed on 31 July 2021).
7. Letters in Applied NanoBioScience, (2020). *Cultivation of Aquatic Plants for Biofiltration of Wastewater.* [online] *10*(1), pp. 1919–1924. Available at: https://nanobioletters.com/wp-content/uploads/2020/10/22846808101.19191924.pdf (accessed on 31 July 2021).

8. Nurul Umairah Mohd Nizam, Marlia Mohd Hanafiah, Izzati Mohd Noor, & Hazwani Izzati Abd Karim, (2020). Efficiency of Five Selected Aquatic Plants in Phytoremediation of Aquaculture Wastewater. [ebook] MDPI. Available at: https://www.mdpi.com/2076–3417/10/8/2712/pdf (accessed on 31 July 2021).
9. Wastewater Treatment, (2021). Floating Aquatic Plant Systems – Wastewater Treatment. [online] Progressive Gardening. Available at: https://www.progressivegardening.com/wastewater-treatment/floating-aquatic-plant-systems.html (accessed on 31 July 2021).
10. Water Environment Federation, (2020). Wastewater Treatment and Water Resource Recovery Facilities (WRRFs) | WBDG – Whole Building Design Guide. [online] Wbdg.org. Available at: https://wbdg.org/resources/wastewater-treatment-and-water-resource-recovery-facilities-wrrfs (accessed on 31 July 2021).
11. Cleantech Water, (2020). What Are the Uses of Small Sewage Treatment Plants? [online] Clean Tech Water. Available at: https://www.cleantechwater.co.in/blog/uses-of-small-sewage-treatment-plants/ (accessed on 31 July 2021).

Chapter

8

Future Economic and Environmental Impacts of Microalgal Technology

CONTENTS

This chapter explains basically the future economic and the environmental impact of the microalgae technology. This chapter also highlights the various aspects of the technology. It further explains the changing global environment and the role of microalgae in this. This chapter also explains the various challenges which are being faced in the current scenario.

It provides a vast description of the role of microalgae and its interrelation with future economics and environmental impacts. This chapter also gives the details regarding the economics on the larger scale. This also highlights the various factors which are mainly influenced by the microalgae.

These factors are explicitly mentioned below and are the elevated carbon dioxide. There are various impacts of the microalgae which are being highlighted with the help of this chapter, and they include the potential environmental impacts, the aquatic impacts, the wastewater treatment, the role of the virus, pathogens, and the parasites.

There are various other impacts that are being clearly observed, and all these factors are mainly highlighted. A detailed description of these factors is being provided here. This chapter not only highlights the impacts of the microalgae in the aquatic areas but also explains the details about the terrestrial impacts and also the terrestrial diversity along with the atmospheric impacts.

8.1. INTRODUCTION

The demand for the water rising capacity is exponentially is seen and there are various combined effects of the growing populations, rising incomes and expanding cities are also being observed A more erratic and uncertain is caused mainly due to the climate change.

Further given the growing demands and the large electricity costs and greenhouse gas emissions are mainly associated with the water supply and waste treatments. A number of countries included waste as a priority sector as there also growing demands for the sanitation as well.

Further in the circular economy, the wastes are mainly valuable resources, and furthermore, the important aspect regarding it is the safe management. The safe management is an efficient investment done in the health of humans and also within the ecosystems.

In the current scenario, it has been observed that the nutrient cycle is basically acting as the matter of concern basically in becoming the critical development and climate strategy for a number of countries. Further, a critical role in the energy fluxes and the nutrient cycles by mainly incorporating and redistributing the dissolved organic matter is being noted and this mainly plays a critical role.

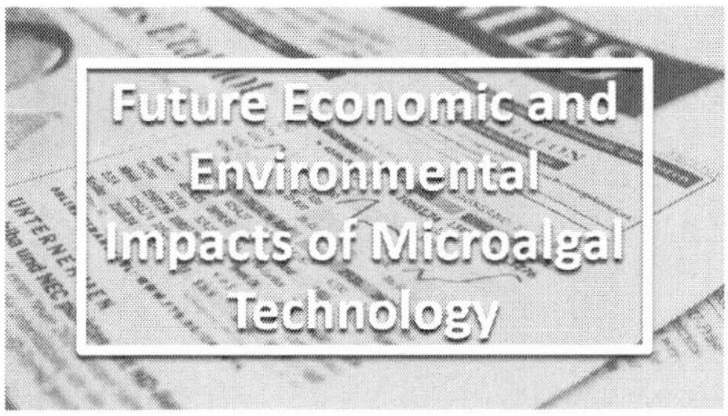

Figure 8.1: The significance of future economic and environmental impacts of microalgal technology.

For the use as the biocatalysts in the environmental biotechnology, they are being observed, and the microorganisms are basically showing a wide range of potential for them. A wide range of biobased chemicals are basically manufactured mainly by the microalgae or the cyanobacteria, and these biochemicals are widely being used.

8.2. THE CHANGING GLOBAL ENVIRONMENT

It has been observed that at the present, the atmospheric CO_2 concentration is much higher when it is compared to the concentration in the geological past, and this shows the changing global environment.

Further, when an example is taken under consideration, changes are being observed such CO_2 levels were mainly 3 to 12 times which was much higher when compared in the present scenario. The above mainly infers truly in the sense that in the present day, the algae are basically living in a low CO_2 environment.

Moreover, currently it has been observed that the CO_2 concentrations in the atmosphere is explicitly increasing and the figures are nearly 280 ppm (28 Pa at sea level) in 1800 to 380 ppm (38 Pa).

However, this atmospheric CO_2 concentration has risen over the last 200 years. Further, at present it came in notice that with most of this increase is basically occurring over the last 100 years.

There is most likely scenario for two and although the above-mentioned forecast varies mainly depends on the inputs for the growth in CO_2 emissions used. The two main scenarios are basically: to attain the values between 750 and 1000 ppm (75–100 Pa) by 2016. The other one is to threefold increase in atmospheric CO_2 concentrations over the next century.

The dissolved carbons and its components are basically discussed in the previous chapters of this book. Furthermore, there are various inevitable consequences of the atmospheric CO_2 increases.

Also, another observation made regarding this is that a proportional increase in dissolved CO_2 in completely air equilibrated is observed in seawater at 15°C, from 14.7 to 38.6 μM and all of this is mainly due to the chemistry of inorganic carbon in water which dictates that an increase in gaseous CO_2 from 38 to 100 pa will lead all of the above-mentioned scenario.

Further, a decrease in the pH is observed and this is in turn the result of the increased CO_2 in the solution. Further, an observation is made that there will be a decrease of (near about 50%) in carbonate concentration due to the equilibria among the dissolved inorganic carbon (DIC) species which are entirely responsible for the above shift.

Figure 8.2: The changing global environment.

This decreased concentration is although observed, but in the case of the bicarbonates, only in minor changes are basically seen and for the total DIC (near about 10% increase).

The alkalinity and the buffering capacity of the waters and as well as on the inputs from the groundwater, are the main components on which the pH and the DIC system will explicitly depend on. Basically, the similar results are obtained for the freshwater systems and these absolute values of the DIC systems and the pH mainly depends on the components.

The last component which is namely the input from the groundwater, is the one that can lead to the freshwater systems, which are in turn being at greater than the atmospheric equilibrium levels of CO_2.

To acquire inorganic carbon under elevated global CO_2, there are some important repercussions for the ability of the aquatic plants which mainly acquire this inorganic carbon. And these important repercussions are basically resulted from the above-mentioned changes in the CO_2 concentrations, but this does not include the HCO_3 concentration.

At present, the global temperatures have been less as for CO_2 levels when estimated in comparing to the paleoclimate which clearly suggests this estimation. As of now when compared to the cretaceous, at least at higher latitudes were considerably higher.

Since the 18th century mainly a significant observation is being made that there is an increase in the temperature, particularly being analyzed for the last years. Mainly before the onset of the industrial revolution, however, the average global temperatures were remarkably constantly for the last 1000 years as well.

Although the actual rise in temperature will not be homogenous and some parts of the world will basically see a far larger shifts in the temperature than in 13 but mainly the increase in atmospheric CO_2 was predicted over the next 100 years is expected to lead to an average further rise in the global sea temperature of 2 to 3°C.

In the current scenario when UVBR and UVCR levels were considered their levels seems to be much higher when compared to near about 2500 to 2000 million years ago. Mainly due to the evolution of oxygenic photosynthesis, the oxygen levels in the atmosphere began to rise at this time.

Further, it has been observed that the formation of ozone layer of sufficient density and depth to significantly attenuate UVBR and UVCR and mainly by 1000 of million years ago and all of this above-mentioned is resulted due to the rise in levels of O_2. Further, approximately present-day fluxes show that the UVR levels are basically reduced.

The ozone levels such as supernovae and further periodic rises in UVR levels could be observed that and all of this could be mainly resulted by the means of UVR fluxes which were essentially constant from 1000 million years ago own wards these fluxes usually created a periodic natural phenomenon which is commonly termed as the bolide impacts, cosmic events such as volcanic eruptions and supernovae.

Basically, these volcanic eruptions and the supernovae are entirely responsible for the above-mentioned impacts of the UVR levels.

Further, it has been estimated that the iron would have been stripped out of the oceans likely if the ferrous ion is UVBR which basically increase the levels of the oxygen and hence this is the complete reason for the iron stripping out also all of this will lead to a greater UVBR penetration.

Lately, it has been observed that mainly due to the decreasing stratospheric ozone concentrations which are likely associated with the release of the chlorofluorocarbons (CFCs) into the atmosphere since the 1970s, the solar flux of UBVR at the Earth's surface is mainly increasing at particular locations.

On the other hand, the increment in the UVBR fluxes to the surface of earth is mostly observed in the regions of Antarctic, and they are marked in the Arctic as well. Over the time period of the last ten years, there is a considerable increment has been observed in UVBR fluxes as far south as France along with the northern Italy. In addition, this has extended over the several regions of the North Atlantic.

In the present interval of time, the UVR penetration into the water bodies is extremely fluctuating. The depth to which UVBR will be reduced to near about 10% surface fluxes that ranges from 0.4 m in the Baltic Sea to more than 16 m in water off the Canary Island. Ecologically, considerable impacts of the UVR have been reported in water as deep as 30 m or even down to 60 to 70 m.

Mainly it is the rapidity of the present global climate change that is of mainly concern and unprecedented in the geological history. In the past decades basically in the past has been mainly exposed to much higher levels of CO_2 concentrations and also a rise in UBVR fluxes was observed and further in some of the latitudes at least at higher temperature.

Further another aspect which came under consideration is that it is somehow crucial to understand how algae will entirely react towards the changes in the global environment which have been predicted and has to be taken over the next century and even beyond this. Mainly the importance of algal photosynthesis in the global carbon budget is a matter of concern.

Further, it has been observed that in the case of photosynthetic and the growth rates there is an increase in CO_2 level is observed which in turn bring out challenges in the cellular composition in the above-mentioned aspects.

A decrease in C:N and increase in C:P and N:P at higher CO_2 concentrations was basically the elemental ratios in a range of microalgae which are probably grown at CO_2 levels. Further, the above-mentioned changes are probably the less marked above the present-day CO_2 levels and they are also very species-dependent.

The mentioned elemental changes were generally examined by Riebesell. Further, in relation to the CO_2 levels the Riebesell's group has basically investigated the various changes in the lipids and fatty acid composition in the coccolithophorid *Emiliania huxleyi*.

The CO_2 concentration was mainly elevated at 1100 fg/cell approximately and also it has been observed that mainly the fatty acids were strongly dependent on this CO_2 concentration which is basically below 14.5 μM.

Moreover, an increase in 14: 0, 18:1 and 18:2 fatty acids was resulted by increasing CO_2 concentration for growth from 1.1 to 14.5 μM. Lately, in the case of the more unsaturated fatty acids (18:5 and 22:6), a strong decline in the percentage content was observed. It has been less obvious that the effects of the CO_2 above the present levels (14 μM).

The effects of the climate changes as the growth conditions were mainly achieved by increasing the total dissolved inorganic carbon but on the other hand the elevated CO_2 levels in the experiments of Riebesell et al.,

Mainly by increasing the alkalinity and keeping the PH is more or less constant. Further, it has been noted that mainly for the microalgal growth and function, basically the possibility of the changes in the macromolecular content under the elevated CO_2 is quite important.

And this is preferred more as compared to just by adjusting the inorganic carbon and the pH. Further, this is not only just for microalgal growth but for also the flows of materials to higher trophic levels.

Urabe et al., 1997 has grown freshwater Daphnia on alga [*Scenedesmus dimorphus* (Turpin) Kützing (= S. acutus)] cultured at a range of CO_2 concentrations there are basically many more examples depict the effects of elevated CO_2 on trophic flow which has been examined.

Rather than direct effect of CO_2 on Daphnia, led to poorer Daphnia growth the under elevated CO_2 Scenedesmus had higher C:P ratios. Further, in the production of the extracellular polysaccharides and this is due to the changing patterns of the macromolecular synthesis. Large amounts of transparent exopolymer particles (TEPs), are mainly produced by *Emiliania huxleyi* blooms and this in turn increase with the cell number and they also produced at the expense of dissolved polysaccharide (PCHO) and this is shown by Engel et al., Further the promotion of sinking of cells is snow was basically led by the TEPs as they are known for the promotion of cell aggregation The elevated CO_2 concentrations up to 30 μM was mainly mentioned in an earlier work Engel showed that TEP production by natural phytoplankton populations in the Baltic Sea was enhanced by its mean.

The phytoplankton aggregation could be increased by the elevated CO_2 and the above observations combined suggested this also a decreasing the recycling of carbon and other nutrients (other nutrients) through the microbial loop mainly by the transport out of the surface mixed layer basically as the marine snow.

Mainly on the nutrient cycling and the ecosystem, such a phenomenon could have a significant effect. Further, in the surface mixed layers of the oceans was mainly functioned by the Science Asia supplement (2006).

The calcification processes will be mainly affected by the lower pH of seawater in a "high CO_2 environment" in addition to the changes associated with the elevated CO_2 per se. The formation of the calcium carbonate in the form of the mineral aragonite was the main factor on which the calcification in the corals and some microalgae are clearly based upon.

Further where the K'sp is the stoichiometric solubility product of the aragonite form of $CaCO_3$ mainly gives the saturation of the seawater with respect to the aragonite. By the partial pressure of CO_2 in solution

$$CO_2 + CO_3^{2-} + H_2O \rightarrow 2HCO_3$$

the aragonite formation is strongly being affected Since Ca^{2+} is essentially constant in seawater, aragonite formation is determined by $[CO_3^{2-}]$. Further, decreased calcification and the decreased $[CO_3^{2-}]$ is the result of the elevated CO_2 levels.

Further, it has been observed that the inhibition of calcification in corals and decreased pH is also one of the consequences of the elevated CO_2 levels. The above-mentioned process is basically quite species dependent.

Lately, it has been observed that the lowered pH than aragonite basically affects less the other forms of the calcium carbonate which is mainly being deposited by the aquatic animals and calcite which is mainly found in some of the macroalgae and most of the coccolithophores. Further, it has been basically observed in the case of coccolithophores that they are significantly affected by the elevated CO_2 levels, and these effects are explicitly being shown.

Later on, under the increased elevated CO_2, the cell size (which was measured as the POC production) by the two species of the coccolithophorid was mainly increased and this was shown basically by Riebesell's work.

Mainly a greater response than *Emilania huxleyi* was shown by Gephyrocapsa oceanic, as compared. The elevated CO_2 is decreased significantly by calcification by the two species. A significant malformation of coccoliths under elevated CO_2 in both the species is being clearly observed.

There is a primary target group which is considered as the children and lactating women and which is identified by the business model. Most severely young children get impacted by the malnutrition, so it is said that through their mothers, it is important to reach this target group, who can be relied on to convey the message of nutrition from peer to peer.

There are some countries where promotion through the conventional media is not suitable, and only in those countries such approach is especially relevant because of its low coverage and high illiteracy rates. It is said that as being the part of the value chain, so to fight against malnutrition, sales activities may contribute indirectly as the additional income can result in improved food intake by both women and their children.

Therefore, it is said that women are also considered as beneficiaries not only who represent sales channels (Heierli and von der Weid, 2007) Product development is known as the process in which it covers the complete process of bringing a new product to the market.

In product development there is always a requirement of the deep understanding of the needs of the customers and in order to develop a functional solution that incorporates critical features into the design of the product there is also the requirement of understanding of the local circumstances.

In its raw form, Spirulina may be sold only to those who are already convinced of its value. It is said that a more attractive product is needed for a wider market. There is an approach to combat malnutrition in developing countries attention and it is for the product development in order to reach targets such as malnourished children are therefore an area that deserves much.

It is said that only with the use of simple technology can be produced and it is also said that the product needs to be easily chewable. There are several examples of specifically designed products meeting these requirements.

India is the country where the Spirulina-enriched "chikkies" found, and they are considered as an excellent example of such product development, which are completely based on local needs and local resources.

There is a snack named as chikki based on peanut and it is a kind of energy biscuit and it also belongs to the most popular snacks for children in India. It is said that spirulina chikkies were very well accepted (Heierli and von der Weid, 2007). So, after this there is another recently marketed product in India takes the form of a candy.

In order to establish a link between malnourished and poor children of India, this product is one of such examples. The retail price for the product is □ 1 or € 0.02, which is an affordable snack for the poor in India. An edible product of this price range is of great potential in the Indian market as it will cater to the poor thus, increasing the chances of working of saleswoman by providing them with a business opportunity.

8.2.1. Global Challenges

The worldwide population is right now experiencing phenomenal development from near about 7 to 9 billion individuals by 2050, which compares to near about 28% expansion (PRB, 2009). In spite of the fact that the pace of population development is in reality decelerating, this halfway advantage is eclipsed by the expanding interest for energy and food production by the fast headway of developing countries (most eminently China and India).

Plainly all countries ought to have an equivalent right to improvement; though, the sheer greatness of aggregate human extension is definitely expanding pressure on future supportability, and a critical reaction is required to address three essential difficulties within recent memory: ensuring against environmental change, an expanding requirement for energy security, and setting up systems to help reasonable populations and financial turn of events.

Consumption of energy and financial development are connected to each other. In spite of the monetary downturn of the worldwide financial crisis in 2008, consumption of energy and Microalgal production frameworks: Global effect of industry scale-up related CO_2 discharges bounced back quickly in 2009 (mostly in developing countries), and in 2010 arrived at a noteworthy high (IEA, 2011b; EIA, 2011). This outlines the solid straight connection between monetary development and energy consumption. Total national output is emphatically connected with worldwide energy consumption (Brown et al., 2011).

It thus, follows that to continue positive financial development into the future as the worldwide population expands, more energy and all the more vigorously proficient procedures for the production of food and fuel will be basic to increment or for sure even keep up expectations for everyday comforts.

In outline, inability to make sure about future energy supplies will contrarily influence financial development and world production. The global network perceives the requirement for feasible energy arrangements regardless of the intricate difficulties presented by the organization of these advances.

The EU Renewable Energy Directive for 10% of sustainable energizes by 2020 (IEA, 2010) and the EU carbon exchanging plan, which is to incorporate flight from 2012 (Ellerman and Buchner, 2007; IEA, 2010), the carbon charge enactment as of late went in Australia at AU$ 23 T–1 C, just as the Renewable Fuel Standard (RFS) in the USA (IEA, 2010; Olmstead and Stavins, 2012), are declaration to this, despite the fact that the universal unification of strategy systems is seemingly as unpredictable as the innovative advancement itself (Metcalf and Weisbach, 2012).

Shockingly, the necessary speed of political change required to accomplish this progress is hampered by falsehood because of personal stakes and a resultant pessimistic open recognition. This is anyway progressively being countered by energy security concerns, which propose that fuel security is an issue not for the removed future however during the current century (Heinberg and Fridley, 2010; IEA, 2010).

Energy securities concerns, for the most part, center upon vaporous and fluid energizes instead of power, and furthermore changes topographically. For instance, right now, flammable gas rules energy security worries in Europe, while oil is the focal point of the US, while Australia (for instance of the Asia Pacific locale), however wealthy in coal and gas, is constrained in oil flexibly and is geologically increasingly sensitive to the expense of fuel importation.

Progressively it is perceived that energy security and atmosphere security are coming into arrangement, and an expansive assorted variety of fuel advances are currently being developed. Therefore, from an environmental change, energy security and monetary point of view, the scale-up of inexhaustible fuel advancements will be progressively significant over the coming decades.

8.2.2. Economics on Large-Scale

8.2.2.1. Microalgal Production Systems

Microalgae have been effectively researched as a feedstock for items, for example, energizes, bio solvents, animal feed, and bio-based plastics and polymers. Microalgae are a promising biomass feedstock because of a wide scope of item prospects, high development rates, and low asset requests.

Despite the fact that there are little to average size business for microalgae production frameworks for high worth human healthful enhancements around the world, enormous scope commercialization of microalgal frameworks for items are restricted by the low financial profits for the items produced.

In view of the measure of 100% worldwide fuel and food production by 2050 that is proportional to near about 0.58 ZJ fuel and near about 0.10 ZJ equivalent of food at incident sun powered light degrees of 20 MJ m^{-2} d^{-1} at PCEs of 2% and 5% would require between 9.7 M km^2 (2% PCE) and 3.9 M km^2 (5% PCE) of devoted microalgal facilities of production.

Whenever incorporated production frameworks are utilized, where food yields establish coproduction from fuel production, at that point, these regions could be lower at 7.5 M km^2 (2% PCE) and 3.0 M km^2 (5% PCE).

If the area required for dedicated systems were to be deployed at a CAPEX cost of US$ 0.15 M ha^{-1} (ignoring economies of scale), a global capital net industry injection of ~US$ 58–145 trillion would be required, while the lesser area for integrated production would require US$ 45–113 trillion (assuming, for simplicity, that a potentially higher cost of capital in integrated systems can be retrieved through increased revenue).

Given that the 2011 global GDP was ~US$ 79 trillion (CIA, 2012), this may seem like an impossible investment. To replace all existing fuel and production, it should however be emphasized that this is the estimated cost and expansion to 2050 levels, and it is based on the above assumptions. On the scale of such an investment as it is use to establish a real perspective, and it is considered as a closer analysis of expenditure on other energy systems is beneficial.

It is said that between 2011 and 2035 (IEA, 2011a), the IEA reported that $38 trillion (in year-2010 dollars) of global investment is required in energy-supply infrastructure. The fossil-fuel industry is considered as another important reference point, which is said to be established technology set, and it already receives over US$ 500 B $yr.^{-1}$ in worldwide subsidy support.

If maintained at this level between 2011 and 2035, this would equate to US$ 12 trillion (or ~US$ 20 trillion to 2050). It is said that 70% of recent major oil discoveries (Tippee et al., 2010) is consider as another important point of reference is that offshore oil discoveries represent and that 'the oil and gas industry will need to invest $50–100 trillion to rebuild its aging infrastructure within the next 7 years to stave off a serious drop in oil and gas production as reported by Dittrick and Izundu (2008).

So, it is said that a rise in oil prices is the magnitude of this expenditure (which would be required in the absence of the large-scale deployment of renewables) which necessitates and it also reinforcing the upward pressure.

Microalgal production systems: It is said that the global impact of industry scale-up sure on prices towards new highs, and it is estimated to be up to ~US$ 200 $bbl.^{-1}$ by 2030 in worst-case scenarios (IEA, 2011c). In addition to it, it must also be considering the capital investment that is currently within the established fossil fuels and associated energy industries (e.g., nuclear and hydro).

Furthermore, it is said that microalgal production is consider as the primary resource restrictions and it will limit, levels of 10% of world production (and possibly above) which are potentially achievable, representing a capital investment of only ~US$ 6–15 trillion.

8.2.2.2. Elevated Carbon Dioxide

Beardall and Raven have studied in detail the performance of growth rate and relative photosynthesis rate of microalgal species that have been affected by elevated carbon dioxide. It is sufficient to say that requirement of inorganic carbon is different in different species as per the activity of carbon dioxide concentrating mechanisms in their cells.

Species depending on diffusive take-up of carbon dioxide are definitely bound to show incitement of photosynthesis and development than species utilizing HCO^{3-} or taking up carbon dioxide effectively by means of a CCM.

Cells lacking CCMs and obtaining carbon dioxide by diffusive take-up are possibly inorganic carbon-constrained under present-day CO_2 levels and may be required to give some incitement of development as barometrical CO_2 levels increment throughout the following 100 years.

One of the most striking instances of this impact of carbon dioxide levels on species lacking CCM action originates from work by Lucy Ball on Lake Windermere and on societies of types of freshwater cryophytes and synurophytes, which have all the earmarks of being the significant exemption to the standard that most microalgae have some type of CCM action.

It is obvious from authentic records that cryophytes and synurophytes possibly show up in noteworthy numbers in Lake Windemere when broken up carbon dioxide levels are at more prominent than air-balance esteems (S. Maberly, Center for Ecology and Hydrology, Lancaster Environment Center, pers. comm.).

Raised barometrical CO_2 may then be relied upon to invigorate the development of these species more than that of different parts of the phytoplankton, however there is no immediate test proof for this at present.

Nonetheless, even in species that have a CCM, contrasts in CCM movement bring about varying affinities for CO2 and somewhat various reactions to carbon dioxide levels, which could prompt moves in species synthesis. For example, in phytoplankton of the Equatorial Pacific, there was no general distinction in complete biomass between tests presented to high carbon dioxide (750 ppm) or low carbon dioxide (150 ppm). Medicines for 8–10 days. In any case, high CO_2 supported the haptophyte Phaeocystis sps. to the detriment of diatoms, while at low CO_2 diatom development was invigorated.

8.3. ENVIRONMENTAL IMPACTS OF MICROALGAE TECHNOLOGY

Figure 8.3: Environmental impacts of microalgae technology.

8.3.1. Aquatic Impacts

Indicators, namely which are the water quality requirements and the water consumption should be necessarily considered when there is a need to evaluate the environmental sustainability of aquatic biomass, and this is particularly based on the cultivation system.

Further there are various examples which are observed of industries investing in large scale microalgae cultivation. Further, it has been observed that on the alga strains mainly the bio-fuels cultivation, water requirements vary completely depended on this Usher et al. (2014).

Moreover, another aspect which is seen is that in order to reduce pressure on natural water resources (i.e., industrial and domestic wastewater), it is possible to use low grade wastes as a water source.

With basically a general outlook, it has been observed that mainly 75% of all wastewater is generated worldwide as sewage is abundant in most of the countries, and thus the collection and treatment methods of this sewage collection varies at different places.

There are various negative impacts mainly on the human health and environment, which are basically the outcomes of the wastewater discharged without treatment into the surface water bodies. For example: There are nearly 9000 wastewater treatment plants are there in UK and this count is on the basis of the domestic water consumption, which is nearly 150 persons per day, which also results in the 120 persons per day of wastewater going into the sewers.

Figure 8.4: Aquatic impacts.

Further the returning to surface water resources is being observed on this basis. As the practice of wastewater management provides low-cost water and nutrients as well as wastewater remediation. Using this domestic and industrial wastewater sources are mainly environmentally and economically beneficial for the large-scale industries. It has been observed that the microalgae mainly thrive by utilizing nutrients which are explicitly available in the water body. Moreover, another quality of microalgae which has been observed is that they are basically highly adaptive to their respective environment.

8.3.2. Wastewater Treatment

Further, it has been clearly observed that as the algae are able to cope with particular pollutants therefore the microalgae cultivation can feasibly be used as a secondary treatment process for various wastewater management.

Further, it has been clearly noted that in the case of humans, microalgae, and animals, a summary of the potential pollutants is basically found in the wastewater and their impacts on these three. Moreover, the chemical use as in standard in wastewater treatment across the world and the need for energy-intensive cleaning processes are reduced mainly by the use of microalgae as a treatment method.

Further to give senses of the extent to which microalgae can be used for the wastewater clean up, the mechanisms for the nutrient removal explicitly depends on the species, and especially for this, they are mainly generalized.

This mechanism moreover helps to deal with the various problems which are being faced. Further, the other important aspect which has been generally noted that basically a well-established technology for wastewater treatments is using microalgae. Moreover, waste stabilization pond systems are one of the popular technologies.

8.4. POTENTIAL ENVIRONMENTAL IMPACTS

Figure 8.5: Potential environmental impacts.

8.4.1. Viruses, Pathogens, and Parasites

Viruses are function as an ecological mechanism for mainly controlling the microalgae populations. Further, it has been observed that the viruses are also affecting microalgae, and in turn, they are thought to be ubiquitous in aquatic environments. Mainly for the large-scale microalgae cultivation, this could in turn lead to two impacts.

This has its various consequences and could results in the loss of the product and also it could knock-on effect on the supply chain for which it is intended and all these results are due to the population collapse outcome of the above scenario.

This population collapse is one hand situation whereas on the other hand it has been observed that these viruses could also be used to control the algal blooms.

Another aspect which explicitly comes in front is that the pathogens will also co-exist with the microalgae. Further, it has been noted that there are high chances that the pathogens may be present in the harvested biomass or in the final process effluent, all of this happens where the water is sourced from the waste streams, particularly municipal or animal waste. And all of this is observed by keeping the facts aside, and the fact regarding this is that in the open ponds, the pathogens can easily be inactivated by the algae cultivation. Further, the end-use of the microalgal product will be adversely affected by this.

It could also lead to potential health risk, or at least it could also lead to the post-treatment and also it might receive before it can be used in any of the product in such risky situations. For the ones those are mainly managing the algal farms, there are also some of the occupational health hazards are being noted.

For example, one such which is mainly identified as the risk to microalgal cultures and the name of this *A. protococcarum*.

Further, in order to protect the microalgal cultures, there is an explicit need to research and also to understand its behavior although several other researches were done regarding this purpose and it founded that the parasite is diverse but to get more clarity related to the aspects further research should be conducted.

8.4.2. Water Footprint (WF)

In the production of the good and services a water footprint is the total amount of the freshwater is mainly embedded. And further, it has been observed that both the surface and the groundwater (blue water footprint) and rainwater (green water footprint) are mainly included in the production of the goods and the services.

The process design and cell biology are the factors on which the evaporation rates, hydraulic retention and also the photosynthetic efficiency explicitly depends. It has been observing that the calculation of the water foot prints is highly sensitive towards the evaporation rates, hydraulic retention time, and also towards the photosynthetic efficiency; therefore, it is necessary to look at these factors and understand them in order to get more clarity.

Further, for example, it can be mainly considered that depending on the local climate 0.48 $m^3\ m^{-1}\ yr^{-1}$ to 2.28 $m^3\ m^{-1}\ yr^{-1}$ in arid regions in arid regions, the evaporation rate from an open system will vary.

Further for microalgae biofuels than other for the biofuels such as soy or palm biodiesel or bioethanol from the sugarcane was mainly found to be lower mainly in the water footprint of a closed photobioreactor for mainly biofuel production was mainly founded. Mainly to freshwater (highest value), the range indicates values from wastewater and seawater (lowest values). Further, it has been noted that the sustainability of both the environmentally and economically of microalgae is mainly confirmed by the wastewater is essential, and this in turn only holds up to the sustainability on these aspects of the biofuels. Mainly on the terms of the nutrient provision and clean water consumption the microalgae biofuels are entirely based upon.

8.5. IMPACTS ON AQUATIC BIODIVERSITY

The controlled eutrophication process is basically defined or knows as the mass cultivation of the microalgae. Further, by the means of the air supply and the regular harvesting, all the above-mentioned needs are being managed well by their means. One of the main risks to the biodiversity is mainly the eutrophication. However, it has been observed the asphyxiation of organisms mainly depends on the oxygen for the process of respiration and asphyxiation is entirely the result of as the decomposition of dead algal biomass consumes the oxygen from the water column. Mainly due to the hypoxia, water toxicity, and turbidity, the impacts of the eutrophication mainly include the reduction in biodiversity. The odorous emissions (e.g., H_2S) and greenhouse gases (e.g., CH_4, CO_2, N_2O) with a strong global warming potential are mainly led due to methane production, which generally occurs in the anaerobic layers. The dominant species can easily replace any organism which depends on the oxygenated water, and these organisms can also be lost. Further, it has been noted that eutrophication events on a large scale, particularly if the cultivation takes place near large water body such as a lake or a coastal area could be resulted if there is any accidental release of water from the cultivation sites into the wider environment.

Figure 8.6: Impacts to the aquatic biodiversity.

Further, the quality of the receiving water body and the size of the release are two factors which are mainly on which the impacts depend. Moreover, for the stabilizing sediments and also for providing the habitats and the food sources for much marine life, the essential elements are mainly the nutrient-rich marine water which also can reduces the seagrass communities and them is one example.

8.6. TERRESTRIAL IMPACTS

The displacement of the food crops for the production, mainly the biofuel production has been met with controversy regarding the displacement. Further, to grow the same quantity of fuel given faster growth rates and higher yields per unit area than the terrestrial crops mainly a key selling point for microalgae biofuels is the reduction of the land needed is basically required.

Further, it has been observed that the predictions based on small scale cultivation were mainly used by many of the initial claims made for the number of biofuels. Mainly in the case of oil production rate of biofuel by estimations is nearly suggested 5775 L ha^{-1} yr^{-1} (4620 L ha^{-1} yr^{-1}) and this is nearly 80% of the conversion efficiency.

Further this information mainly infers that this conversion is significantly lower than the other published estimates. Further, equivalent to the 48% of the current US petroleum imports for transportation which is mainly based on the consumption in 2011, the microalgae has is potential to generate 220 $\times$ 109 L yr^{-1} of oil, and this is mainly suggested by a study mainly under the current technology.

Figure 8.7: Terrestrial impacts of microalgae.

Further, it has been estimated that for biodiesel from corn, 594 M ha for biodiesel from soybean nearly that the replace is nearly 50% of the US transport fuels, and also 1540 ha of land would be needed for this. Although it has also been noted that around 43 M ha for biodiesel from microalgae.

8.6.1. Contamination and Leaks

Further for the designers there are many designs. Also, at lower cost, the open ponds basically allow the large-scale cultivation. There are mainly being made vulnerable to contamination by the means of the open design.

By making the cultural conditions unfavorable to the native species and the risk can also be minimized by the means of altering culture conditions. Further, it has been noticing that a threat to the direct environment from leaching of the pond's contents in to the grounds can be caused if the ponds are not correctly designed or constructed, and these two are major reasons behind the above-mentioned threats.

The losses of toxicants where microalgae are also being used as wastewater treatment facility and the salinization in situations where the marine algae are cultivated on the land are mainly the two examples of the above scenario or condition.

The contamination of groundwater could be leaded by the means of whilst content of the ponds. These contents of the pond might not be necessarily being toxic. The components which basically allow the light to penetrate through the microalgae are namely the photobioreactors (PBRs) are translucent containers.

It has been lately observed that PBRs are basically less susceptible to the contamination are they are closed. A significant impact mainly could be seen from the leaks of the containers, and this explicitly depends on the volume. The example for them is that when they can be nearly located near a natural source of water. The above context mainly infers that it would be therefore easy to rectify and also potentially easy to detect.

8.6.2. Impacts to Terrestrial Diversity

Mainly by the destruction of the habitat, the displacement of local fauna can be caused. This is, therefore, the outcomes of the constructions of ponds. To assess the level of the impact of destruction of the large-scale ponds would have could easily be done by the use of the environmental impact assessment surveys. Further, it has been observed by the NCR that mainly due to the infrastructure development for the algal biofuels has severely resulted in the effects on the terrestrial biodiversity from changing the landscape pattern.

Figure 8.8: Impacts of microalgae to the terrestrial diversity.

Basically, reservoir construction acts as an excellent example which shows that how large water projects has mainly affected the local biodiversity. Further, it also provides a guide as to what some of the changes could be as whilst the size of the reservoir construction is generally greater than the expected change from the microalgal ponds.

8.6.3. Atmospheric Impacts

Mainly to the water and land systems are basically the areas where whilst the direct impacts of the microalgae cultivation is the most apparent one. Further, it has been observed in case of the atmosphere is also impacted mainly by the means of the large-scale microalgae cultivation, which possess a potential.

Mainly the type of the cultivation system acts as the major factor on which the scale of impact depends. The potential to reduce the greenhouse gas emissions by the microalgae cultivation uptake during cultivation is also observed.

Moreover, this section at potential gaseous and aerosol emissions from microalgae cultivation is also seen. Further, as a result of pollutant species are emitted, the above context shows that the direct impacts and atmospheric reaction can take place and provide the result.

8.7. WASTE TO ENERGY

The Micro algae-based biofuels basically produce the energy efficiency ratio (EER), which is relatively lower than that of the oil bearing crops, and basically, the waste to energy indicates this. The use of inorganic source of nitrogen, operation and maintenance of photobioreactor and harvesting/dewatering of microalgae biomass are mainly are the uses of production, basically the energy hotspots were identified in the overall algal biofuels production process and the process are the above ones.

It should be noted that for culturing microalgae for biofuel production should mainly be coupled along with the wastewater treatment and it was basically highly recommended. Further, when the main aim was taking into the consideration was that to minimize the heavy dependency on the organic nutrients, to reduce the water footprints and the carbon also is aimed at treating the wastewaters.

Mainly for the treatment of the industrial, agricultural, wastewater and municipal are the areas where the high-rate algal ponds (HRAP) are used for treating. Further, it has been noted that the high-rate algal ponds are shallow.

The high-rate algal lakes (HRAL) are shallow and they have been generally utilized for the treatment of metropolitan, modern, and horticultural wastewaters. In 1957, Golueke et al., first proposed the utilization of high-rate algal ponds for enormous scope creation of algal based biofuels utilizing wastewaters.

High-rate algal ponds have a profundity of 0.2 to 1.0 meter and blending is given by paddle wheel. The paddlewheel gives a level water speed 0.15 m/s to 0.3 m/s. The CO_2 can be included into a countercurrent gas sparging siphon (near about 1.5 m profundity) making violent stream inside the lake.

The raceway lakes can be arranged as different circles around focal partitioning dividers. These raceway lakes can be perplexed to upgrade blending. Algal biomass creation from wastewater high-rate algal ponds offers an appealing recommendation with respect to the carbon impression.

The high-rate algal ponds development and activity are required for giving wastewater treatment. Thus, the algal biomass is a result which speaks to the biofuels feedstock which is liberated from natural weight. For instance, Shilton et al., gather feed of paddlewheel that baffle flow raceway lakes with numerous circles and confuses.

Researchers gave a model for a town of 25,000 individuals in the English countryside; they were utilizing a lake for rewarding their wastewater rather than electrochemical treatment framework bringing about sparing 35 million kWh over a 30-year structure life. In UK, a study showed that for every kWh of electricity 0.43 kg of carbon dioxide is produced. So, in order to overcome the effect of 500 tons of CO_2 produced there will be 200 hectares of pine forest will be required.

8.8. CONCLUSION

In the conclusion of this chapter, the marine microalgae and its role in the future economics and the environmental impacts of it states that as in the current scenario, microalgae is being highly popularized due to its high demand and various advantages. The role of microalgae in the changing global environment and the way it has been impacting our current global conditions.

It has been mainly observed that the level of carbon dioxide is currently rising and the role of microalgae in this rise. The rise in the level of the carbon dioxide and the marine microalgae are somewhat inter related. There are various economic aspects of the microalgae and how they are economy of a place can rise by the help of microalgae.

Slowly and steadily the micro algae have gained pace, and the demand of the microalgae within the market is rising. There are various environmental impacts of the marine microalgae, including both the aquatic and the terrestrial impacts of the micro algae. The atmosphere is also being impacted by the microalgae. These impacts are clearly being observed, and they are also study in order to get more and more clarity regarding it.

REFERENCES

1. Usher, P., Ross, A., Camargo-Valero, M., Tomlin, A., & Gale, W., (2014). An overview of the potential environmental impacts of large-scale microalgae cultivation. *Biofuels*, [online], *5*(3), 331–349. Available at: https://www.tandfonline.com/doi/full/10.1080/17597269.2014.913925 (accessed on 31 July 2021).
2. Amar Doshi, (2017). Economic Analyses of Microalgae Biofuels and Policy Implications in Australia. [ebook] p.1. Available at: http://eprints.qut.edu.au/103532/1/Amar_Doshi_Thesis.pdf (accessed on 31 July 2021).
3. Clarens, A., Nassau, H., Resurreccion, E., White, M., & Colosi, L., (2011). Environmental Impacts of Algae-Derived Biodiesel and Bioelectricity for Transportation. *Environmental Science & Technology,* [online] *45*(17), 7554–7560. Available at: https://www.researchgate.net/publication/51507067_Environmental_Impacts_of_Algae-Derived_Biodiesel_and_Bioelectricity_for_Transportation (accessed on 31 July 2021).
4. Jegathese, S., & Farid, M., (2014). Microalgae as a Renewable Source of Energy: A Niche Opportunity. *Journal of Renewable Energy*, [online] 2014, pp. 1–10. Available at: https://www.hindawi.com/journals/jre/2014/430203/ (accessed on 31 July 2021).
5. Khan, M., Shin, J., & Kim, J., (2018). The promising future of microalgae: current status, challenges, and optimization of a sustainable and renewable industry for biofuels, feed, and other products. *Microbial Cell Factories,* [online] *17*(1). Available at: https://www.ncbi.nlm.nih.gov/pmc/articles/PMC5836383/ (accessed on 31 July 2021).
6. Simas-Rodrigues, C., Villela, H., Martins, A., Marques, L., Colepicolo, P., & Tonon, A., (2015). Microalgae for economic applications: advantages and perspectives for bioethanol. *Journal of Experimental Botany,* [online] *66*(14), 4097–4108. Available at: https://www.researchgate.net/publication/275657250_Microalgae_for_economic_applications_Advantages_and_perspectives_for_bioethanol (accessed on 31 July 2021).
7. Usher, P., Ross, A., Camargo-Valero, M., Tomlin, A., & Gale, W., (2014). An overview of the potential environmental impacts of large-scale microalgae cultivation. *Biofuels*, [online] *5*(3), 331–349. Available at: https://www.researchgate.net/publication/271672957_An_Overview_of_the_Potential_Environmental_Impacts_of_Large_Scale_Microalgae_Cultivation (accessed on 31 July 2021).

8. Simas-Rodrigues, C., Villela, H., Martins, A., Marques, L., Colepicolo, P., & Tonon, A., (2015). Microalgae for economic applications: advantages and perspectives for bioethanol. Journal of Experimental *Botany*, [online] *66*(14), 4097–4108. Available at: https://academic.oup.com/jxb/article/66/14/4097/2893380 (accessed on 31 July 2021).
9. Clarens, A., Nassau, H., Resurreccion, E., White, M., & Colosi, L., (2011). Environmental Impacts of Algae-Derived Biodiesel and Bioelectricity for Transportation. *Environmental Science & Technology,* [online] *45*(17), 7554–7560. Available at: https://pubs.acs.org/doi/abs/10.1021/es200760n (accessed on 31 July 2021).